Studies in Systems, Decision and Control

Volume 328

Series Editor

Janusz Kacprzyk, Systems Research Institute, Polish Academy of Sciences, Warsaw, Poland

The series "Studies in Systems, Decision and Control" (SSDC) covers both new developments and advances, as well as the state of the art, in the various areas of broadly perceived systems, decision making and control–quickly, up to date and with a high quality. The intent is to cover the theory, applications, and perspectives on the state of the art and future developments relevant to systems, decision making, control, complex processes and related areas, as embedded in the fields of engineering, computer science, physics, economics, social and life sciences, as well as the paradigms and methodologies behind them. The series contains monographs, textbooks, lecture notes and edited volumes in systems, decision making and control spanning the areas of Cyber-Physical Systems, Autonomous Systems, Sensor Networks, Control Systems, Energy Systems, Automotive Systems, Biological Systems, Vehicular Networking and Connected Vehicles, Aerospace Systems, Automation, Manufacturing, Smart Grids, Nonlinear Systems, Power Systems, Robotics, Social Systems, Economic Systems and other. Of particular value to both the contributors and the readership are the short publication timeframe and the world-wide distribution and exposure which enable both a wide and rapid dissemination of research output.

Indexed by SCOPUS, DBLP, WTI Frankfurt eG, zbMATH, SCImago.

All books published in the series are submitted for consideration in Web of Science.

More information about this series at http://www.springer.com/series/13304

Rafael Martínez-Guerra ·
Fidel Meléndez-Vázquez ·
Iván Trejo-Zúñiga

Fault-tolerant Control and Diagnosis for Integer and Fractional-order Systems

Fundamentals of Fractional Calculus and Differential Algebra with Real-Time Applications

Springer

Rafael Martínez-Guerra
Departamento de Control Automático
CINVESTAV-IPN
Ciudad de México, Mexico

Fidel Meléndez-Vázquez
Departamento de Física y Matemáticas
Universidad Iberoamericana
Ciudad de México, Mexico

Iván Trejo-Zúñiga
División de Mecatrónica y TIC
Universidad Tecnológica
de San Juan del Río
San Juan del Río, Querétaro, Mexico

ISSN 2198-4182   ISSN 2198-4190   (electronic)
Studies in Systems, Decision and Control
ISBN 978-3-030-62096-7   ISBN 978-3-030-62094-3   (eBook)
https://doi.org/10.1007/978-3-030-62094-3

This Springer imprint is published by the registered company Springer Nature Switzerland AG
The registered company address is: Gewerbestrasse 11, 6330 Cham, Switzerland

To the memory of my father
Carlos Martínez Rosales.

To my wife and sons
Marilen, Rafael and Juan Carlos.

To my mother and brothers
Virginia, Victor, Arturo, Carlos, Javier and
Marisela.

Mes remerciements particuliers au Dr.
Michel Fliess, ainsi qu'à mon ami le Dr.
Sette Diop.

—Rafael Martínez-Guerra

In loving memory of my father,
Fidel Meléndez García, and my grandmother,
Teresa Vázquez Castillo.

To my mother María Victoria,
my brother Diego Alejandro
and my nephew Alejandro.

To the memory of my grandmother Teresa.

—Fidel Meléndez-Vázquez

To my parents and brothers
Carmen, Rogelio, Omar and Rogelio Jr.

To my girl
Karla Y.

To my friends and students.

—Iván Trejo-Zúñiga

Preface

This book is about algebraic and differential methods, as well as fractional calculus, applied to diagnose and reject faults in nonlinear systems, which are of integer or fractional order. This represents an extension of a very important and widely studied problem in control theory, namely, fault diagnosis and rejection (using differential algebraic approaches), to systems presenting fractional dynamics, i.e., systems whose dynamics are represented by derivatives and integrals of non-integer order; these kind of systems are having a lot of attention recently due to the real-world applications they have and also due to the search for generalization of representing dynamical systems.

This book is self-contained; this means that the basic concepts and tools of mathematics required to understand the methodologies proposed are explained in former chapters. The intended audience of the book are practicing engineers, mathematicians, physicists and, in general, to researchers in all areas with a minimum working knowledge of mathematics. However, it also contains advanced research topics for people interested in the area of states and faults estimation. For this, we apply tools and methods from differential algebra and fractional calculus.

The plan of the book is as follows. Chapter 1 presents a brief overview of the main themes covered, giving an introduction and the state of the art on both fault diagnosis and fault-tolerant control, as well as on differential algebra and fractional-order calculus and dynamical systems. Also mentioned are some recent works comprising these areas. In Chap. 2 some definitions from differential algebra are briefly introduced, which is the main mathematical tool used for fault diagnosis, and also for the design of the fault-tolerant control scheme. Some concepts are presented, such as differential field extensions, the differential primitive element, nonlinear dynamics, and canonical forms. Chapter 3 deals with the problem of fault diagnosis for a certain class of nonlinear systems based on a differential algebraic approach, using the left invertibility condition. The detection of faults in the system is based on input-output measurements; the outputs are the signals measured from sensors, and their number is important in order to know if the system is diagnosable or not, i.e., if the faults can be diagnosed from its available information. Then, in order to reconstruct the faults on the system, an invariant observer is proposed which uses the property of invariance with respect to transformation groups. Finally, another nonlinear observer is used for fault diagnosis, for comparison purposes. In Chap. 4, the proposed methodology for fault diagnosis and fault-tolerant control for integer-order systems is developed. The class of system considered is presented, along with the notions of algebraic observability and the reduced-order observer for fault diagnosis. Later, a canonical form is defined from the nominal system, and by means of the output tracking error dynamics, a high-gain observer is constructed; from the dynamics of this observer the fault-tolerant dynamical controller is obtained. It is demonstrated that the closed-loop system is stable. Finally, the proposed methodology is applied in an academic example and in the model of the three-tank system Amira DTS200.

Chapter 5 presents some basic concepts of fractional calculus, as well as of the theory of fractional-order dynamical systems. Concepts such as Gamma and Mittag-Leffler functions are presented, together with the fractional-order integrals and derivatives. Next, commensurate-order fractional systems are defined. Later, some controllers developed for fractional-order dynamical systems are presented, as well as some existing stability results for linear and nonlinear fractional-order systems. In Chap. 6, a new fault diagnosis scheme applied to fractional order nonlinear systems affected by unknown signals is developed. This diagnostic methodology is based on a new fractional reduced-order observer capable of estimating multiple faults and state variables. The strategy of reconstruction is carried out employing two new definitions: The fractional algebraic observability and the fractional algebraic diagnosability, these definitions allow to know beforehand whether faults, state variables, or both can be estimated. Finally, some numerical simulations are performed and discussed in order to show the effectiveness and versatility of the suggested approach. In Chap. 7, a fractional integral reduced-order observer is defined, which is shown to be Mittag-Leffler stable. Its design is based on the fractional algebraic observability property and is used to deal with the synchronization and anti-synchronization problems in commensurate and incommensurate-order fractional chaotic systems by means of the master-slave configuration. The observer (slave) estimates the states of the master system and synchronizes with it. It is also used to estimate some fractional derivatives of the output that appear in the dynamics. This methodology is applied in the fractional Lorenz chaotic system with commensurate dynamics, i.e., where the fractional order of the dynamics of all the states is the same, and in the fractional Rössler chaotic system with incommensurate dynamics. After applying the scheme, simulations are performed in order to obtain numerical results. In Chap. 8, the methodology proposed in Chap. 4 for fault diagnosis and fault-tolerant control is extended to commensurate-order fractional systems, and a fractional-order reduced-order observer is proposed. The stability results presented in Chap. 4 are used in the error linear dynamics and the stability proof of the closed-loop system, which in this case is verified to be Mittag-Leffler stable. In order to prove the methodology developed for fractional-order systems, simulations are performed in the fractional-order models of the Van der Pol oscillator and a DC motor. By last, a comparison of the results obtained with the DC motor is made with the simulation results obtained with the same system but with integer-order dynamics, for different fractional orders. In Chap. 9, a fractional sliding mode controller based on a robust PI^α observer for fractional systems with bounded disturbances is presented. This controller is introduced to achieve closed-loop Mittag-Leffler stability for uncertain fractional systems; meanwhile, the robust PI^α observer is designed such that a $\mathscr{L}_2$ stability is obtained via quadratic Lyapunov functions and the tuning is achieved with linear matrix inequality approach. The performance and stability of the proposed controller are analyzed and demonstrated through a generalized $n\alpha$-differentiator. Finally, to illustrate the effectiveness of the proposed approach a comparative analysis against the fractional Super-Twisting is carried out.

México City, México
June 2020

Rafael Martínez-Guerra
Fidel Meléndez-Vázquez
Ivan Trejo-Zúñiga

Contents

Acronyms and Symbols

CFD	Caputo fractional derivative
FAD	Fractional algebraic diagnosability
FAO	Fractional algebraic observability
FD	Fault diagnosis
FGOCF	Fractional generalized observability canonical form
FHGO	Fractional high-gain observer
FIROO	Fractional (proportional) integral reduced-order observer
FMGOCF	Fractional multi-input multi-output generalized observability canonical form
FNLS	Fractional nonlinear systems
FROO	Fractional (proportional) reduced-order observer
FO	Fractional-order
FODE	Fractional-order ordinary differential equation
FOS	Fractional-order systems
FSTA	Fractional super-twisting algorithm
FTC	Fault-tolerant control
GFCF	Generalized fractional canonical form
GLFD	Grünwald-Letnikov fractional derivative
GOCF	Generalized observability canonical form
HGO	High-gain observer
IFAO	Incommensurate fractional algebraic observability
IO	Integer-order
MMGOCF	Multi-input multi-output generalized observability canonical form
OT	Output tracking
PIROO	Proportional integral reduced-order observer
PROO	Proportional reduced-order observer
RFD	Robust fractional differentiator
RLFD	Riemann-Liouville fractional derivative

RLFI	Riemann-Liouville fractional integral
ROO	Reduced-order observer
SMC	Sliding mode control
SMO	Sliding mode observer
STA	Super-twisting algorithm
$\forall$	For all
$\in$	Belongs to
$\subset$	Subset of
$\rightarrow$	Tends to
$\mapsto$	Maps to
$\sum$	Summation
$\mathbb{R}$	The set of real numbers
$\mathbb{R}^+$	Positive real numbers
$\mathbb{R}^{m \times n}$	The set of all $m \times n$ matrices with elements from $\mathbb{R}$
$\mathbb{R}^{m \times n}_+$	The set of all $m \times n$ matrices with elements from $\mathbb{R}^+$
A^{-1}	Inverse of matrix A
max	Maximum
min	Minimum
sup	Supremum, the least upper bound
inf	Infimum, the greatest lower bound
$f : S_1 \rightarrow S_2$	A function f mapping a set S_1 into a set S_2
$\lambda_{max}(P)(\lambda_{min}(P))$	The maximum (minimum) eigenvalue of a symmetric matrix P
$P > 0$	A positive definite matrix P
$\dot{y}$	The first derivative of y with respect to time
$\ddot{y}$	The second derivative of y with respect to time
$\dddot{y}$	The third derivative of y with respect to time
$y^{(i)}$	The ith derivative of y with respect to time
lim	Limit
$\|a\|$	Absolute value of a scalar a
$\|\|x\|\|$	The Euclidean norm of a vector
$B_R(a)$	Ball of radius R centered in a: $\{x\,\|\,\|\|x - a\|\| \leq R, R \in \mathbb{R}\}$
difftrd°	Differential transcendence degree
∞	Infinity
	Designation of the end of proofs
$K\langle u \rangle$	Differential field generated by the field K, the input $u(t)$ and the time derivative of u
$K\langle u, y \rangle$	Differential field generated by the field K, the input $u(t)$ the output $y(t)$ and the time derivatives of u and y
$\widehat{x}$	Estimation of variable x
$^{RL}I^{(\alpha)}x(t)$	Riemann-Liouville fractional integral of $x(t)$ order α
$^{RL}D^{(\alpha)}x(t)$	Riemann-Liouville fractional derivative of $x(t)$ of order α
$^{C}D^{(\alpha)}x(t)$	Caputo fractional derivative of $x(t)$ of order α

$D^{(\alpha)}x(t)$	Caputo fractional derivative of $x(t)$ of order α
$n!$	Factorial of n
$\Gamma(z)$	Gamma function of z
$B(p,q)$	Beta function with parameters p, q
$E_\alpha(z)$	Mittag-Leffler function of z with parameter α
$E_{\alpha,\beta}(z)$	Mittag-Leffler function of z with parameters α, β

List of Figures

Chapter 1
Overview

As it has been stated, this book is about algebraic and differential methods, as well as fractional calculus, applied to diagnose and reject faults in nonlinear systems, which are of integer or fractional order. Thus, in this chapter, a brief overview of the main themes involved is given.

1.1 Fault Diagnosis and Fault-Tolerant Control

The level of automation present in human life has reached nowadays a very high level, both in industry and in daily life. Still, the number of tasks taken by computers is growing every day: aeroplanes, cars, robots, biomedical applications and numerous others. However, not in all of them the possible appearance of faults in actuators, sensors and other components has been considered during the design process.

Faults are signals that appear in different kind of physical systems of main interest in industrial and engineering applications. Faults are undesired deviations of one or more characteristics or parameters of a system with respect to their nominal operation. Due to this, faults have detrimental effects on the correct performance and good conditions of physical equipment, which may result in problems such as economic dropout, system damage or even catastrophe.

Hence, guaranteeing the system safety and reliability becomes a critical issue on the design of automatic systems. For this purpose, it is imperative to design and implement methods to detect the appearance of these signals and eliminate or diminish their consequences as early as possible. Motivated by these facts, the problems of fault diagnosis and fault-tolerant control have been studied and applied in engineering systems design for more than three decades.

© The Author(s), under exclusive license to Springer Nature Switzerland AG 2021
R. Martínez-Guerra et al., *Fault-tolerant Control and Diagnosis for Integer
and Fractional-order Systems*, Studies in Systems, Decision and Control 328,
https://doi.org/10.1007/978-3-030-62094-3_1

The action of determining in binary form the existence of faults, as well as the moment when they occur, is known as fault detection. If it is also specified the kind of faults and their location, the process is known as fault isolation. Moreover, if the magnitude and behaviour with respect to time of faults is determined, such action is called fault identification. Finally, when detection, isolation and identification are performed simultaneously, it has been performed fault diagnosis.

Fault diagnosis (FD) has been a research area of great interest for a long time. There exist several works that study this problem with the aid of control systems, such as [1, 48], which involves residual generation, decoupling of undesired inputs, and adaptive approaches. For linear systems, the geometric approach [25] has also been applied, where the concept of unobservability subspaces is used together with residual generation.

There exist also several works that deal with FD in nonlinear systems [13, 14, 49]. The geometric approach for fault detection and isolation via residual generation has also been extended to nonlinear systems, where distributions are considered as the nonlinear equivalents of the unobservability subspaces [8].

As mentioned formerly, once the faults are diagnosed, it is imperative to implement a control strategy to eliminate or diminish to the extent possible their effects. This is why the need of a fault-tolerant control arises, which is designed according to the system in question, that eliminates the effects caused by the faults in the systems once they are diagnosed.

Fault-tolerant control (FTC) forms an essential part of many applications found in automation and engineering [5, 19, 23, 42]. There exist also several approaches to obtain such kind of control; i.e., the study [30] provides a revision of basic literature that covers most areas of FTC. The book [4] presents a model-based approach for FTC. Particularly, in the field of nonlinear systems, several favorable results have been obtained by applying algebraic techniques [10, 16, 17, 19, 23].

1.2 Differential Algebra

Differential algebra was introduced in 1932 by the American mathematician Ritt [35, 36]. This subject contributed tools to the theory of differential equations like those proposed in commutative algebra for algebraic equations, i.e. the theory of systems of differential equations which are algebraic in the unknowns and their derivatives, completed by the theory of systems of algebraic equations.

Abstract algebra matured thanks to R. Dedekind, L. Kronecker, D. Hilbert, E. Noether, et al. The first treatise on the subject was then being published by van der Waerden [46, 47]. The basics of differential algebra presented here may be found in the work of Kolchin [20], although the work is difficult for someone with a limited working knowledge of algebra.

The theory of fields was created in the second half of the nineteenth century in order to avoid lengthy manipulations of algebraic equations. In the same way, the modern language of differential geometry, of great current use in nonlinear control [15], was partly invented to escape from the "debauch" of indices.

1.3 Fractional Calculus and Fractional-Order Systems

In its basic definition, fractional calculus is the mathematical generalization of classical calculus, which involves integrals and derivatives of non-integer order. This theory was born historically on September 30, 1695, relatively close to the invention of classical calculus. In a letter addressed to G. W. Leibniz, G. de L'Hôpital questions him, regarding his recently proposed notation d^n/dx^n, about what would happen if $n = 1/2$, to which Leibniz answered that it would lead to a paradox from which someday useful consequences would be obtained. The etymology of the theory is due to this fact, but in a strict sense fractional calculus involves integrals and derivatives of real order, let it be rational or irrational, or even complex; recently, the order of these operators may be constant, time-variant, or even random or diffuse.

Leibniz's comment led to the appearance of integrals and derivatives of an arbitrary order in an almost definitive form at the end of the 19th century, mainly due to J. Liouville, A. K. Grünwald, A. V. Letnikov and G. F. B. Riemann. However, for almost three centuries it developed mainly as a pure theoretical field, useful only to mathematicians.

It wasn't until the seventies of the 20th century that fractional calculus began to be the object of several conferences and specialized treatises. As for the first conference, the merit is due to B. Ross who, after his doctoral dissertation on the area, organized the First Conference on Fractional Calculus and its Applications in 1974 [38]. For the first monograph the merit is attributed to K. B. Oldham and I. Spanier who, after a joint collaboration begun in 1968, published a book devoted to the subject in 1974 [29].

Nowadays there exist a great number of books [9, 27, 28, 31, 33], as well as scientific papers and specialized conferences, focused either on the theoretical part or in the real world applications.

Fractional-order dynamical systems, i.e., systems which mathematical model is represented by means of derivatives and integrals of non-integer order, have been strongly studied in recent decades. This is due to the great amount of applications and multidisciplinary physical problems that present dynamics with fractional derivatives and integers, such as material science [7], thermal systems [11], damped mechanical systems [12], electrical circuits [18], bioengineering [22], polimeric behaviour [26], diffusion problems [29], electromagnetism [37], finance [41], viscoelasticity [43], electromechanics [50], etc. Furthermore, sometimes fractional equations give better approximations of system behaviour than the integer-order ones.

Applications of fractional calculus have been also developed in control theory. For example, A. Oustaloup studied fractional-order algorithms for control of dynamical systems and demonstrated the superiority of the CRONE (non-integer order robust control) [40] over the PID controller. Furthermore, I. Podlubny proposed a generalization of the PID controller, called the $PI^\lambda D^\mu$ controller [34], with a λ-order integrator and a μ-order differentiator. From here, other control techniques have been extended to fractional orders, such as the sliding mode control, the model-reference adaptive control and the reset control [28]. In a similar way, several stability results have been developed based on several approaches [21, 24, 39].

Particularly, regarding FD in fractional systems, approaches such as residual generators [2], the generalized space of dynamic parity [3] and sliding modes [32] have been used. As for FTC, [6], robust control against actuator faults [44] and the $PI^\lambda D^\mu$ [45] have been used.

1.4 Scope of the Book

In this book, the problems of FD and FTC are addressed by means of numerical simulations and experimental results in classic integer-order systems, with emphasis in FD. For this end, some observers are introduced.

Furthermore, the methodology proposed for integer systems is then extended to tackle with FD and FTC in fractional-order systems, again with emphasis in FD. Also for this end, different fractional-order observers are presented, while showing their utility in other applications.

Finally, it is worth to note that in this book the problem of FD, both for integer and fractional-order, is restricted to the case of faults in the actuators, i.e., when the faults appear in the state equations, coupled with the control inputs.

$$\dot{x} = g(x, u, f)$$
$$y = h(x).$$

References

1. Alcorta-García, E., Frank, P.: Deterministic nonlinear observer-based approaches to fault diagnosis: a survey. Control Eng. Pract. **5**(5), 663–670 (1997)
2. Aribi, A., Aoun, M., Farges, C., Najar, S., Melchior, P., Abdelkrim, M.N.: Generalized fractional observers scheme to fault detection and isolation. In: 10th International Multi-Conference on Systems, Signals & Devices (SSD), Hammamet, Tunisia, March 18–21, pp. 1–7 (2013)
3. Aribi, A., Farges, C., Aoun, M., Melchior, P., Najar, S., Abdelkrim, M.N.: Fault detection based on fractional order models: application to diagnosis of thermal systems. Commun. Nonlinear Sci. **19**(10), 3679–3693 (2014)
4. Blanke, M., Kinnaert, M., Lunze, J., Staroswiecki, M.: Diagnosis and Fault-Tolerant Control. Springer, Berlin (2003)

5. Chen, L., Liu, S.: Fault diagnosis and fault-tolerant control for a nonlinear electro-hydraulic system. In: Proceedings of the Conference on Control and Fault Tolerant Systems, Nice, France, October 6–8, pp. 269–274 (2010)
6. Chouki, R., Aribi, A., Aoun, M., Abdelkrim, M.N.: Additive fault tolerant control for fractional order model systems. In: Proceedings of the 16th International Conference on Sciences and Techniques of Automatic Control & Computer Engineering - STA'2015, Monastir, Tunisia, December 21–23, pp. 340–345 (2015)
7. De Espíndola, J.J., da Silva Neto, J.M., Lopes, E.M.O.: A generalized fractional derivative approach to viscoelastic material properties measurement. Appl. Math. Comput. **164**(2), 493–506 (2005)
8. De Persis, C., Isidori, A.: A geometric approach to nonlinear fault detection and isolation. IEEE Trans. Autom. Control **46**(6), 853–865 (2001)
9. Diethelm, K.: The Analysis of Fractional Differential Equations: An Application-Oriented Exposition Using Differential Operators of Caputo Type. Springer, Berlin (2010)
10. Fliess, M., Join, C., Sira-Ramírez, H.: Closed-Loop Fault-Tolerant Control for Uncertain Nonlinear Systems. In: Meurer, T., et al. (eds.) Control and Observer Design LNCIS 322, pp. 217–233. Springer, Berlin (2005)
11. Gabano, J.D., Poinot, T.: Fractional modelling and identification of thermal systems. Signal Process **91**(3), 531–541 (2011)
12. Gaul, L., Klein, P., Kempfle, S.: Damping description involving fractional operators. Mech. Syst. Signal Process **5**(2), 81–88 (1991)
13. Hammouri, H., Kinnaert, M., El Yaagoubi, E.H.: Observer based approach to fault detection and isolation for nonlinear systems. IEEE Trans. Autom. Control **44**(10), 1879–1884 (1999)
14. Hutter, F., Dearden, R.: Efficient on-line fault diagnosis for nonlinear systems. In: Proceedings of the 7th International Symposium on Artificial Intelligence, Robotics and Automation in Space (2003)
15. Isidori, A.: Nonlinear Control Systems. Springer, New York (1989)
16. Join, C., Ponsart, J.C., Sauter, D., Theilliol, D.: Nonlinear filter design for fault diagnosis: application to the three-tank system. IEE Proc.-Control Theory Appl. **152**(1), 55–64 (2005)
17. Join, C., Sira-Ramírez, H., Fliess, M.: Control of an uncertain three tank system via on-line parameter identification and fault detection. In: Proceedings of 16th Triennial World IFAC Conference, Prague, Czech Republic, July 2005, vol. 38, no. 1, pp. 251–256 (2005)
18. Kaczorek, T., Rogowski, K.: Fractional Linear Systems and Electrical Circuits. Springer, Cham (2015)
19. Kiltz, L., Join, C., Mboup, M., Rudolph, J.: Fault-tolerant control based on algebraic derivative estimation applied on a magnetically supported plate. Control Eng. Pract. **26**(2014), 107–115 (2014)
20. Kolchin, E.R.: Differential Algebra and Algebraic Groups. Academic, New York (1973)
21. Li, Y., Chen, Y., Podlubny, I.: Stability of fractional-order nonlinear dynamic systems: Lyapunov direct method and generalized Mittag-Leffler stability. Comput. Math. Appl. **59**(5), 1810–1821 (2010)
22. Magin, R.: Fractional Calculus in Bioengineering. Begell House, Redding (2006)
23. Mai, P., Hillermeier, C.: Fault-tolerant tracking control for nonlinear systems based on derivative estimation. In: Proceedings of the American Control Conference, Baltimore, MD, USA, June 30–July 2, pp. 6486–6493 (2010)
24. Matignon, D.: Stability results for fractional differential equations with applications to control processing. Ma Comput. Sci. Eng. **2**(1996), 963–968 (1996)
25. Massoumnia, M.A., Verghese, G.C., Willsky, A.S.: Failure detection and identification. IEEE Trans. Autom. Control **34**(3), 316–321 (1989)
26. Metzler, R., Schick, W., Kilian, H.G., Nonnenmacher, T.F.: Relaxation in filled polymers: a fractional calculus approach. J. Chem. Phys. **103**(16), 7180–7186 (1995)
27. Miller, K.S., Ross, B.: An Introduction to the Fractional Calculus and Fractional Differential Equations. Wiley, New York (2010)

28. Monje, C.A., Chen, Y., Vinagre, B.M., Xue, D., Feliu, V.: Fractional-Order Systems and Controls: Fundamentals and Applications. Springer, London (2010)
29. Oldham, K.B., Spanier, J.: The Fractional Calculus: Theory and Applications of Differentiation and Integration to Arbitrary Order. Academic, New York (1974)
30. Patton, R.J.: Fault-tolerant control systems: the 1997 situation. In: IFAC symposium on fault detection supervision and safety for technical processes, vol. 3, pp. 1033–1054 (1997)
31. Petráš, I.: Fractional-Order Nonlinear Systems: Modeling. Analysis and Simulation. Springer, Beijing (2011)
32. Pisano, A., Usai, E.: Second-order sliding mode approaches to disturbance estimation and fault detection in fractional-order systems. In: Proceedings of the 18th IFAC World Congress, Milano, Italy, August 28–September 2, pp. 1033–1054 (2011)
33. Podlubny, I.: Fractional Differential Equations: An Introduction to Fractional Derivatives, Fractional Differential Equations, to Methods of their Solution and Some of their Applications. Academic, San Diego (1999)
34. Podlubny, I.: Fractional-order systems and $PI^\lambda D^\mu$-controllers. IEEE Trans. Autom. Control **44**(1), 208–214 (1999)
35. Ritt, J.F.: Differential Equations from the Algebraic Standpoint. American Mathematical Society, New York (1932)
36. Ritt, J.F.: Differential Algebra. American Mathematical Society, New York (1950)
37. Rosales, J.J., Gómez, J.F., Guía, M., Tkach, V.I.: Fractional electromagnetic waves. In: Proceedings of the LFNM*2011 International Conference on Laser & Fiber-Optical Networks Modeling, Kharkov, Ukraine, September 4–8, pp. 1–3 (2011)
38. Ross, B. (ed.): Fractional Calculus and Its Applications. In: Proceedings of the International Conference held at the University of New Haven, June 1974. Springer, Berlin (1975)
39. Sabatier, J., Farges, C., Trigeassou, J.C.: A stability test for non-commensurate fractional order systems. Syst. Control Lett. **62**(9), 739–746 (2013)
40. Sabatier, J., Oustaloup, A., García-Iturricha, A., Lanusse, P.: CRONE Control: principles and extension to time-variant plants with asymptotically constant coefficients. Nonlinear Dyn. **29**(1), 363–385 (2002)
41. Scalas, E., Gorenflo, R., Mainardi, F.: Fractional calculus and continuous-time finance. Phys. A **284**(14), 376–384 (2000)
42. Seron, M.M., De Don, J.A., Richter, J.H.: Integrated sensor and actuator fault-tolerant control. Int. J. Control **86**(4), 689–708 (2013)
43. Shaw, S., Warby, M.K., Whiteman, J.R.: A comparison of hereditary integral and internal variable approaches to numerical linear solid elasticity. In: Proceedings of the XIII Polish Conference on Computer Methods in Mechanics, Pozna Poland, May 5–8, 1997 (1997)
44. Shen, H., Song, X., Wang, Z.: Robust fault-tolerant control of uncertain fractional-order systems against actuator faults. IET Control Theory A **7**(9), 1233–1241 (2013)
45. Talange, D., Joshi, S.: Fractional order fault tolerant controller for AUV. In: Proceedings of the 18th International Conference on Automatic Control, Modelling & Simulation (ACMOS '16), Venice, Italy, January 29–31, 2016, pp. 287–292 (2016)
46. Van der Waerden, B.L.: Moderne Algebra, vol. I. Springer, Berlin (1930)
47. Van der Waerden, B.L.: Moderne Algebra, vol. II. Springer, Berlin (1931)
48. Willsky, A.: A survey of design methods in observer-based fault detection systems. Automatica **1**(2), 601–611 (1976)
49. Xu, A., Zhang, Q.: Nonlinear system fault diagnosis based on adaptive estimation. Automatica **40**(7), 1181–1193 (2004)
50. Yu, W., Luo, Y., Pi, Y.: Fractional order modeling and control for permanent magnet synchronous motor velocity servo system. Mechatronics **23**(7), 813–820 (2013)

Chapter 2
Fundamentals of Differential Algebra

In this chapter the basics of differential algebra are briefly presented, which is a mathematical tool used in the design of controllers for linear and nonlinear systems. The concepts presented here will be used in the next chapters for fault diagnosis and to build the fault-tolerant control scheme in the integer and fractional-order cases.

2.1 Differential Rings

Definition 2.1 Let A be a commutative ring with unit element. The derivative (differentiation) ∂ of A is a closed mapping $\partial : A \to A$ such that all pairs $(a, b), a, b \in A$, satisfying

$$+ : A \times A \to A$$
$$(a, b) \mapsto a + b$$

and

$$\cdot : A \times A \to A$$
$$(a, b) \mapsto a \cdot b$$

also satisfy

$$\partial(a + b) = \partial a + \partial b$$

and

$$\partial(a \cdot b) = (\partial a)b + a(\partial b)$$

© The Author(s), under exclusive license to Springer Nature Switzerland AG 2021
R. Martínez-Guerra et al., *Fault-tolerant Control and Diagnosis for Integer and Fractional-order Systems*, Studies in Systems, Decision and Control 328, https://doi.org/10.1007/978-3-030-62094-3_2

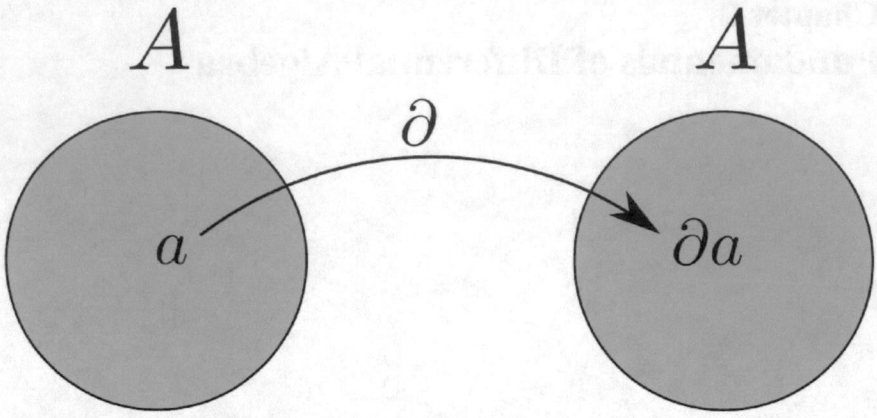

Fig. 2.1 Derivation in a commutative ring A

where $a + b$, $a \cdot b \in A$ (Fig. 2.1).

Definition 2.2 A differential ring A is a commutative ring with a unit element that is provided with a finite set of differentiation operators over A such that $\forall \partial_1, \partial_2 \in \Delta$, $\forall a \in A$

$$\partial_1 \partial_2 a = \partial_2, \partial_1 a$$

Definition 2.3 A differential ring is called ordinary or partial if the finite set Δ respectively contains a single or more than one differentiation operator.

Differential rings are useful in the study of ordinary algebraic differential equations and partial algebraic differential equations, respectively.

Definition 2.4 A differential subring A' of a differential ring A is a subring of A such that with the differential operator in Δ, A' is a differential ring.

Definition 2.5 A constant a of a differential ring is an element such that $\forall \partial \in \Delta$, $\partial(a) = 0$.

Corollary 2.1 *The set of constants of a differential ring is a differential subring.*

2.2 Fields and Differential Fields

2.2.1 Fields

Definition 2.6 Let K be a field. A subfield K' of K is a subring of K provided with the internal composition laws (addition and multiplication) induced from those of K. This gives a structure to the field, and K is called a field extension of K'.

Definition 2.7 Let L be a field extension of the field K and let S be a subset of L. The subfield generated by $K \cup S$ is represented by $K(S)$, and it is called the subfield generated by the adjunction of elements of S to K.

If S is reduced to a single element $\alpha \in L$ (and $\alpha \notin K$), then $K(\alpha)$ is called a simple extension.

Definition 2.8 An element of L is said to be algebraic over K if and only if it satisfies an algebraic equation with coefficients in K. The extension L/K is algebraic if and only if any element of L is algebraic over K.

Example 2.1 Consider the field $K = \mathbb{Q}$, and let $\sqrt[n]{p} \in L = \mathbb{R}$. Then $\sqrt[n]{p}$ is algebraic over $\mathbb{Q}$, since there exists a polynomial, namely $P(x) = x^n - p$, such that $P(\sqrt[n]{p}) = 0$.

Example 2.2 Every element of $L = \mathbb{Q}$ is algebraic over $K = \mathbb{Z}$.

Example 2.3 Consider $i \in \mathbb{C}$. Then i is algebraic over $\mathbb{R}$, since i satisfies the polynomial $P(x) = x^2 + 1$, namely, $P(i) = 0$.

Definition 2.9 An element $a \in L$ is said to be transcendental over K if and only if it is not algebraic over K. This implies the nonexistence of a single-variable polynomial $p(x)$ over K such that $p(a) = 0$. The extension L/K is said to be transcendental if and only if there exists at least one element of L which is transcendental over K.

Example 2.4 The extension $\mathbb{R}/\mathbb{Q}$ is transcendental since e and Π are transcendental over $\mathbb{Q}$.

There exist, for transcendental extensions, concepts which are nonlinear analogs of dimension and basis for a vector space.

Definition 2.10 A set $\{\xi_i \mid i \in I\}$ of elements in L is said to be K-algebraically dependent if and only if there exists at least one polynomial $P(x_1, \ldots, x_v)$ over K, such that $P(\xi_{i_1}, \ldots, \xi_{i_v}) = 0$. A set which is not K-algebraically dependent is said to be K-algebraically independent.

Definition 2.11 A K-algebraically independent set which is maximal with respect to inclusion is called a transcendence basis of L/K. The cardinality of this set is called the transcendence degree of L/K, which is noted as $\operatorname{tr} d^\circ L/K$.

Therefore, an extension L/K is algebraic if and only if $\operatorname{tr} d^\circ L/K = 0$.

Example 2.5 Let $K = \mathbb{R}$, and consider the field of rational functions in a single indeterminate s (one variable), denoted by $\mathbb{R}(s)$:

$$f(s) = \frac{a_0 + a_1 s + \cdots + a_n s^n}{b_0 + b_1 s + \cdots + b_m s^m}$$

where $a_1, b_i \in \mathbb{R}$. Hence, if s is transcendental over $K = \mathbb{R}$, then $\operatorname{tr} d^\circ \mathbb{R}(s)/\mathbb{R} = 1$.

Definition 2.12 Let L be an extension of the field K. One says that the extension L/K is finite if there exists a finite subset x of L such that $L = K(x)$.

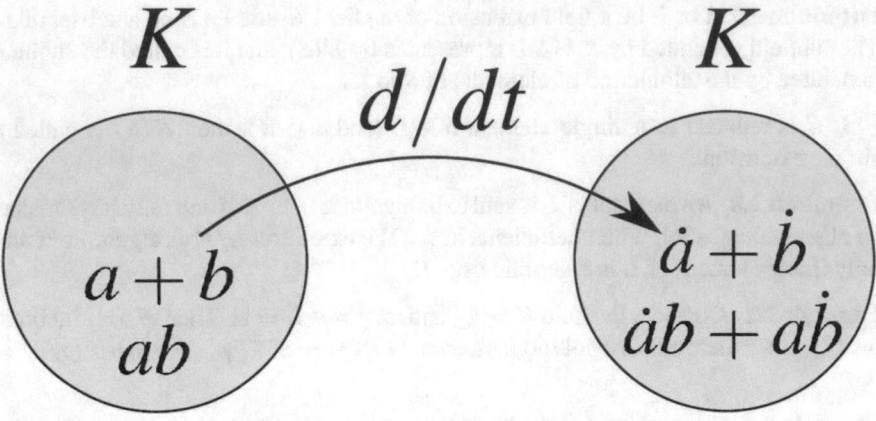

Fig. 2.2 Differential field K

2.2.2 Differential Fields

Definition 2.13 A differential field K is a commutative field which is equipped with a single derivation $d/dt = $ " $\cdot$ ". This derivation obeys the usual rules

$$\frac{d}{dt}(a + b) = \dot{a} + \dot{b}$$

$$\frac{d}{dt}(ab) = \dot{a}b + a\dot{b}$$

$\forall\, a, b \in K$ (Fig. 2.2).

Definition 2.14 A constant of K is an element $c \in K$ such that $\dot{c} = 0$.

Corollary 2.2 *The set of constants of a differential field is a differential subfield.*

Definition 2.15 A differential field extension L/K is given by two differential fields K, L such that:

(i) K is a subfield of L,
(ii) the derivation of K is the restriction to K of the derivation of L (Fig. 2.3).

Definition 2.16 The intersection of a set of differential subfields of L is also a differential subfield of L.

Definition 2.17 An element $\xi \in L$ is said to be differentially algebraic over K if and only if it satisfies a differential equation $P(\xi, \dot{\xi}, \dots, \xi^{\alpha}) = 0$, where P is a polynomial over K in ξ and its derivatives. The equation $P = 0$ is called an algebraic differential equation. The extension L/K is said to be differentially algebraic if and only if any element of L is differentially algebraic over K.

Fig. 2.3 Differential field
extension L/K

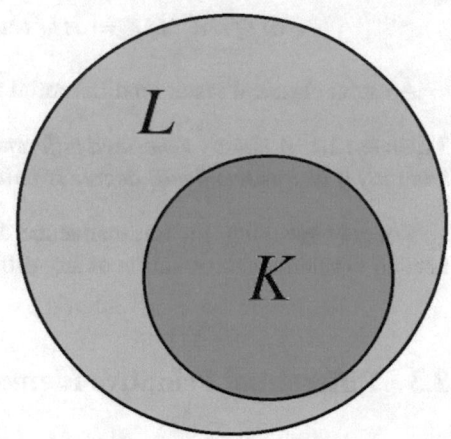

Example 2.6 Let $K = \mathbb{Q}$, then $a = e^t \in L$ satisfies the differential equation $\dot{x} - x = 0$.

Example 2.7 Let $K = \mathbb{R}$. Let L be the field $\mathbb{R}(x)$, and $\dot{x} = e^x, \ddot{x} = \dot{x}e^x$. The element x satisfies the differential equation with coefficients in $\mathbb{R}$ given by $\ddot{x} - \dot{x}^2 = 0$.

Definition 2.18 An element $a \in L$ is said to be differentially transcendental over K if and only if it is not differentially algebraic over K. This means that no algebraic differential equation over K exists, which is satisfied by a. The extension L/K is said to be differentially transcendental if and only if there exists at least one element of L which is differentially transcendental over K.

Definition 2.19 A set $\{\xi_i \mid i \in I\}$ of elements in L is said to be differentially K-algebraically dependent if and only if the set of derivatives of any order $\{\xi_i^{\nu_i} \mid i \in I, \; \nu_i = 0, 1, 2, \ldots\}$ is K-algebraically dependent. In other words, the elements $\{\xi_i\}$ satisfy some algebraic differential equation. A set which is not differentially K-algebraically dependent is said to be differentially K-algebraically independent.

Definition 2.20 A set of differentially K-algebraically independent elements which is maximal with respect to inclusion is called a differential transcendence basis of L/K. The cardinality of this set is called the differential transcendence degree of L/K, which is noted as diff tr $d°L/K$.

Therefore, the extension L/K is differentially algebraic if and only if diff tr $d°L/K = 0$.

Furthermore, let $u = \{u_i \mid i \in I\}$ be a differential transcendence basis of L/K. Note as $K\langle u \rangle$ the differential field generated by K and the elements of u. The differential fields L and $K\langle u \rangle$ do not, in general, coincide but the extension $L/K\langle u \rangle$ is differentially algebraic.

Let $K \subseteq L \subseteq M$ be a tower of differential field extensions. Then

$$diff trd°M/K = diff trd°M/L + diff trd°L/K$$

A useful classical result on differential algebra is stated as follows.

Theorem 2.1 *A finitely generated differential extension is differentially algebraic if and only if its transcendence degree is finite.*

Roughly speaking, the transcendence degree is the number of initial conditions needed to calculate the solution of any differential algebraic equation.

2.3 Differential Primitive Element

Take a finitely generated algebraic extension L/K. The theorem of the primitive element states the following.

Theorem 2.2 *There exists a single element $\gamma \in L$, which is a primitive element, such that $L = K(\gamma)$; i.e., L is generated by K and γ.*

Now, take a finitely generated differentially algebraic extension L/K, and assume that K is not a field of constants. The theorem of the differential primitive element states that [3]:

Theorem 2.3 *There exists a single element $\delta \in L$, which is a differential primitive element, such that $L = K\langle\delta\rangle$; i.e., L is generated by K and δ.*

2.4 Algebraic Approach of Nonlinear Dynamics

Definition 2.21 Let k be a given differential ground field. Let $k\langle u\rangle$ be the differential field generated by k and the elements of a finite set $u = (u_1, \ldots, u_m)$ of differential quantities. A dynamics is a finitely generated differentially algebraic extension $K/k\langle u\rangle$. The input u is said to be independent if and only if u is a differential transcendence basis of K/k.

Let n be the transcendence degree of $K/k\langle u\rangle$. Take a finite set $\xi = (\xi_1, \ldots, \xi_v)$, $v \geq n$, of elements in K, which contains a transcendence basis of $K/k\langle u\rangle$.

Each of the derivatives $\dot{\xi}_1, \ldots, \dot{\xi}_v$ are $k\langle u\rangle$-algebraically dependent over ξ:

$$L_1(\dot{\xi}_1, \xi, u, \dot{u}, \ldots, u^{(\alpha)}) = 0$$

$$\vdots$$

$$L_v(\dot{\xi}_v, \xi, u, \dot{u}, \ldots, u^{(\alpha)}) = 0$$

where $L_1, \ldots, L_v$ are polynomials over k. We should stress that the above differential equations are implicit. Assume that the variables take real (or complex) values. When the Jacobian matrix

$$\begin{pmatrix} \partial L_1/\partial \dot{\xi}_1 & & (0) \\ & \ddots & \\ (0) & & \partial L_v/\partial \dot{\xi}_v \end{pmatrix}$$

has full rank v, we get from the implicit function theorem explicit differential equations:

$$\dot{\xi}_1 = a_1(\xi, u, \dot{u}, \ldots, u^{(\alpha)}) = 0$$

$$\vdots$$

$$\dot{\xi}_v = a_v(\xi, u, \dot{u}, \ldots, u^{(\alpha)}) = 0$$

The explicit form is only locally valid, i.e., in domains where the Jacobian matrix has full rank.

The variable ξ is known as a generalized state or a state for short. The integer v is called its dimension. A minimal (generalized) state, i.e., a state of minimal dimension, is a transcendence basis of $K/k\langle u \rangle$; its dimension is n. Such a state is characterized by the $k\langle u \rangle$-algebraic independence of its components.

Take two minimal states $x = (x_1, \ldots, x_n)$, $\bar{x} = (\bar{x}_1, \ldots, \bar{x}_n)$. Any component of $\bar{x}$ is $k\langle u \rangle$-algebraically dependent over the components of x. There must exist polynomials $P_1, \ldots, P_n$ over k such that:

$$P_1(\bar{x}_1, x, u, \dot{u}, \ldots, u^{(\alpha)}) = 0$$

$$\vdots$$

$$P_n(\bar{x}_n, x, u, \dot{u}, \ldots, u^{(\alpha)}) = 0$$

Also, because (by minimality) the components of x are $k\langle u \rangle$-algebraically dependent over those of x, one can show that the Jacobian matrix

$$\begin{pmatrix} \partial P_1/\partial \bar{x}_1 & & (0) \\ & \ddots & \\ (0) & & \partial P_n/\partial \bar{x}_n \end{pmatrix}$$

must have full rank.

The preceding discussion can be summarized as follows.

Definition 2.22 Let $K/k\langle u \rangle$ be a dynamics, i.e., a finitely generated differentially algebraic extension. A state representation is, in general, implicit and can only be put in explicit form locally. A minimal state is a transcendence basis of $K/k\langle u \rangle$. Two minimal states are related by equations involving the control variables and a finite number of their derivatives.

Remark 2.1 In what follows, we will assume that every algebraic differential extension L/K is generated by a finite set $(\xi_1, \ldots, \xi_n, u_1, \ldots, u_m)$. Therefore, the linear combination

$$\sum_i^n \alpha_i \xi_i + \sum_j^m \beta_j u_j, \qquad \alpha_i, \beta_j \in K\langle u \rangle$$

is a differential primitive element.

2.5 Canonical Forms

2.5.1 Generalized Controller Canonical Form

In order to apply the theorem of the differential primitive element to the nonlinear dynamics $K/k\langle u \rangle$, we have to assume that $k\langle u \rangle$ is not a field of constants. This is indeed the case where, for example, the set u of control variables is nonempty and independent.

Let δ be a differential primitive element of $K/k\langle u \rangle$. Consider the sequence of derivatives $\delta, \dot{\delta}, \ldots, \delta^{(v)}, \ldots$.

Lemma 2.1 ([2]) *The set* $(\delta, \dot{\delta}, \ldots, \delta^{(v)})$ *is* $k\langle u \rangle$-U)-*algebraically independent (respectively, dependent) if and only if* $v \leq n - 1$ *(respectively,* $v \geq n$*), where* $n = \operatorname{tr} d^\circ K/k\langle u \rangle$.

Corollary 2.3 ([2]) $(\delta, \dot{\delta}, \ldots, \delta^{(n-1)})$ *is a transcendence basis of* $K/k\langle u \rangle$.

We write

$$C(\delta, \dot{\delta}, \ldots, \delta^{(n)}, u, \dot{u}, \ldots, u^{(\alpha)}) = 0$$

where C is a polynomial over k. The selection $x_1 = \delta, x_2 = \dot{\delta}, \ldots, x_n = \delta^{(n-1)}$ is what we called a minimal state of the dynamics $K/k\langle u \rangle$, which yields the equations

$$\dot{x}_1 = x_2$$
$$\vdots$$
$$\dot{x}_{n-1} = x_n$$
$$C(x_1, x_2, \ldots, x_n, \dot{x}_n, u, \dot{u}, \ldots, u^{(\alpha)}) = 0$$

This is what we call a generalized global controller canonical form. We may locally obtain a generalized local controller canonical form:

$$\dot{x}_1 = x_2$$

$$\vdots$$

$$\dot{x}_{n-1} = x_n$$

$$\dot{x}_n = c\,(x_1, x_2, \ldots, x_n, u, \dot{u}, \ldots, u^{(\alpha)})$$

In the context of dynamical systems, x represents the state variable and u is the input or control variable.

2.5.2 Generalized Observability Canonical Form

Let $x_i = y^{(i-1)}, i = 1, \ldots, n$, where y represents the output variable. We obtain the following local state space, which may be seen as a generalization of the observability canonical form [1].

$$\dot{x}_1 = x_2$$

$$\vdots$$

$$\dot{x}_{n-1} = x_n$$

$$\dot{x}_n = C\,(x_1, x_2, \ldots, x_n, u, \dot{u}, \ldots, u^{(\alpha)})$$

$$y = x_1$$

In comparison to the usual canonical forms for linear systems, the canonical forms presented here depend on the input variable and its time derivatives; hence the name "generalized". Some other examples may be seen in [4].

References

1. Fliess, M.: Automatique et corps différentiels. Forum Math. **1**, 227–238 (1989)
2. Fliess, M.: Generalized controller canonical forms for linear and nonlinear dynamics. IEEE Trans. Autom. Control **35**(9), 994–1001 (1990)
3. Kolchin, E.R.: Differential Algebra and Algebraic Groups. Academic, New York (1973)
4. Martínez-Guerra, R., Cruz-Ancona, C.D.: Algorithms of Estimation for Nonlinear Systems: A Differential and Algebraic Viewpoint. Springer, Cham (2017)

Chapter 3
Fault Diagnosis by Means of Invariant Observers

The level of automation has reached a high level, both, in industry and in daily life. Still, the number of tasks taken by computers is growing every day; in airplanes, biomedical applications, cars, robots, and numerous others. Only in few of them possible faults, in e.g. actuators, sensors and components have been considered during the design. However, in most applications they are not considered. In such systems, faults or abnormal changes of individual parts can occur and result in economic dropout, system damage or even catastrophe. Hence, guaranteeing the system safety and reliability becomes a critical issue on the design of automatic systems. For this purpose, the most important thing is to detect the faults in systems as early as possible. Motivated by these facts, the systems diagnosis has been studied for more than three decades, many papers dealing with this problem can be found (for instance [1, 9]). For the case of nonlinear systems several approaches have been proposed [1]. On the other hand, for the fault diagnosis problem, exist an alternative approach based on an algebraic and differential framework [3, 7, 8].

This chapter deals with the diagnosis of nonlinear systems, with the main goal of detecting malfunctions in the system based on input-output measurements. The outputs are mainly signals measured from sensors, their number is important in order to know if the system is diagnosable or not. In this chapter, the diagnosis problem is tackled as a left invertibility problem throughout the concept of differential output rank ρ that guarantees the system diagnosability. For comparison purposes two schemes of observers are proposed in order to estimate the fault signals, one of them is the so-called reduced-order observer based on a free-model approach [8], the second one is an invariant observer based on a Generalized Observability Canonical Form and the property of invariance.

© The Author(s), under exclusive license to Springer Nature Switzerland AG 2021
R. Martínez-Guerra et al., *Fault-tolerant Control and Diagnosis for Integer and Fractional-order Systems*, Studies in Systems, Decision and Control 328,
https://doi.org/10.1007/978-3-030-62094-3_3

3.1   The Left Invertibility Condition

Some definitions of differential field extensions and the concept of differential transcendent degree are required to follow this section, as well as, the notion of diagnosability of the system. Further details can be found in [6, 8] and references therein.

Let us consider the class of nonlinear systems with faults described by the following relationship

$$\dot{x} = A(x, \bar{u}) \tag{3.1}$$

$$y = h(x, \bar{u}) \tag{3.2}$$

where $x = (x_1, \ldots, x_n)^T \in \mathbb{R}^n$ is a state vector, $u = (u_1, \ldots, u_m)^T \in \mathbb{R}^m$ is a know input vector (possibly a designed control vector), $f = (f_1, \ldots, f_\mu)^T \in \mathbb{R}^\mu$ is an unknown input vector (called fault vector), $\bar{u} = (u, f)^T \in \mathbb{R}^{m+\mu}$ and $y = (y_1, \ldots, y_p) \in \mathbb{R}^p$ is the measured output vector. A and h are assumed to be analytic functions.

In order to ease the reading of the chapter some definitions about the differential output range of a system will be presented.

Definition 3.1 The differential output rank ρ of a system is equal to the differential transcendence degree of the differential extension $K \langle y \rangle$ over the differential field K, i.e.,

$$\rho = diff \ tr \ d^\circ K \langle y \rangle / K. \tag{3.3}$$

Property 3.1 ([5]) *Let K, L and M be differential fields such that $K \subset L \subset M$. Then*

$$diff \ tr \ d^\circ(M/K) = diff \ tr \ d^\circ(M/L) + diff \ tr \ d^\circ(L/K). \tag{3.4}$$

Property 3.2 *The differential output rank ρ of a system is smaller or equal to $min(m, p)$, i.e., $\rho = diff \ tr \ d^\circ K \langle y \rangle / K \leq min(m, p)$, where m and p are the total number of inputs and outputs, respectively.*

Proof A proof of Property 3.2 can be given in the following manner: an input-output system, with input $u = (u_1, \ldots, u_m)$ and output $y = (y_1, \ldots, y_p)$ is defined by the next conditions:

- $(u_1, \ldots, u_m)$ are differentially K-algebraically independent i.e;

$$diff \ tr \ d^\circ K \langle u \rangle / K = m \tag{3.5}$$

- $(y_1, \ldots, y_p)$ are differentially algebraic over $K \langle u \rangle$, i.e., $K \langle u, y \rangle / K \langle u \rangle$ is differentially algebraic or

$$diff \ tr \ d^\circ K \langle u, y \rangle / K = 0 \tag{3.6}$$

Now, consider the field tower

$$K \subset K\langle u \rangle \subset K\langle u, y \rangle \tag{3.7}$$

by Property 3.1

$$diff\ tr\ d^\circ K\langle u, y \rangle / K = diff\ tr\ d^\circ K\langle u, y \rangle / K\langle u \rangle + diff\ tr\ d^\circ K\langle u \rangle / K \tag{3.8}$$

Replace Eqs. (3.5) and (3.6) into Eq. (3.8), we obtain

$$diff\ tr\ d^\circ K\langle u, y \rangle / K = m \tag{3.9}$$

Now, consider the field tower

$$K \subset K\langle y \rangle \subset K\langle u, y \rangle \tag{3.10}$$

using Property 3.1

$$diff\ tr\ d^\circ K\langle u, y \rangle / K = diff\ tr\ d^\circ K\langle u, y \rangle / K\langle y \rangle + diff\ tr\ d^\circ K\langle y \rangle / K \tag{3.11}$$

Substituting the result showed in Eq. (3.9) into Eq. (3.11)

$$m = diff\ tr\ d^\circ K\langle u, y \rangle / K\langle y \rangle + diff\ tr\ d^\circ K\langle y \rangle / K \tag{3.12}$$

Since the differential transcendence degree is not negative, we have that

$$\rho = diff\ tr\ d^\circ K\langle y \rangle / K \leq m \tag{3.13}$$

In a similar manner; $y = (y_1, \ldots, y_p)$ and $\rho = diff\ tr\ d^\circ K\langle y \rangle / K \leq p$
Finally,

$$\rho = diff\ tr\ d^\circ K\langle y \rangle / K \leq \min(m, p). \tag{3.14}$$

$\square$

Remark 3.1 Roughly speaking the differential output rank ρ is also the maximum number of outputs that are related by a differential polynomial equation with coefficients over K (independent of x and u).

A practical way to determinate the differential output rank is by taking into account all possible differential polynomials of the form

$$h_r(y_1, \ldots, y_p) = 0 \tag{3.15}$$

If it is possible to find r independent relations of the form (3.15), then the differential output rank is given by $\rho = p - r$ that is to say, there exists only $p - r$ independent outputs.

Proposition 3.1 ([4]) *Consider a class of systems given by* (3.1). *A system is said to be left invertible if and only if*

$$\rho = diff\ tr\ d^\circ K\langle y\rangle/K = diff\ tr\ d^\circ K\langle u, f\rangle/K. \tag{3.16}$$

Property 3.1 is the main tool used to prove the following theorem that looks quite natural. The theorem shows the relationship between the diagnosability and the left invertibility condition.

Theorem 3.1 *If the system* (3.1) *is left invertible, the fault vector f can be obtained by means of the output vector.*

Proof Consider the following field tower

$$K \subset K\langle u\rangle \subset K\langle u, f\rangle \subset K\langle u, y, f\rangle \tag{3.17}$$

$$K \subset K\langle y\rangle \subset K\langle u, y\rangle \subset K\langle u, y, f\rangle \tag{3.18}$$

from Eq. (3.17) and Property 3.1, we have

$$\begin{aligned}
diff\ tr\ d^\circ K\langle u, y, f\rangle/K &= diff\ tr\ d^\circ K\langle u, y, f\rangle/K\langle u, f\rangle \\
&\quad + diff\ tr\ d^\circ K\langle u, f\rangle/K\langle u\rangle + diff\ tr\ d^\circ K\langle u\rangle/K \\
&= 0 + \mu + m \\
&= m + \mu
\end{aligned} \tag{3.19}$$

From Proposition 3.1 and $diff\ tr\ d^\circ K\langle y\rangle/K = m + \mu$, from (3.18) we obtain

$$\begin{aligned}
diff\ tr\ d^\circ K\langle u, y, f\rangle/K &= diff\ tr\ d^\circ K\langle u, y, f\rangle/K\langle u, y\rangle \\
&\quad + diff\ tr\ d^\circ K\langle u, y\rangle/K\langle y\rangle + diff\ tr\ d^\circ K\langle y\rangle/K \\
&= diff\ tr\ d^\circ K\langle u, y, f\rangle/K\langle u, y\rangle \\
&\quad + diff\ tr\ d^\circ K\langle u, y\rangle/K\langle y\rangle + m + \mu
\end{aligned} \tag{3.20}$$

From Eqs. (3.19) and (3.20) we have

$$diff\ tr\ d^\circ K\langle u, y, f\rangle/K\langle u, y\rangle + diff\ tr\ d^\circ K\langle u, y\rangle/K\langle y\rangle + m + \mu = m + \mu$$

this implies that

$$diff\ tr\ d^\circ K\langle u, y, f\rangle/K\langle u, y\rangle = -diff\ tr\ d^\circ K\langle u, y\rangle/K\langle y\rangle \tag{3.21}$$

Since the transcendence degree is always positive, we have the following:

$$diff\ tr\ d°K\langle u, y, f\rangle / K\langle u, y\rangle = 0 \tag{3.22}$$

This means that f is differentially algebraic over $K\langle u, y\rangle$. Thus, the diagnosability condition is satisfied and the theorem is proven. □

Example 3.1 Consider the system

$$\dot{x}_1 = x_2 + f_1 + f_2 \tag{3.23}$$
$$\dot{x}_2 = x_1 x_2 + f_1 \tag{3.24}$$
$$y_1 = x_1 \tag{3.25}$$
$$y_2 = x_2 \tag{3.26}$$

the differential output rank of the system (3.23)–(3.26) is $\rho = 2$ since

$$h_r = (y_1, y_2) = 0 \tag{3.27}$$

In fact, given that ρ is equal to the number of faults, it can be concluded that the above system (3.23)–(3.26) is left invertible, in other words, f_1 and f_2 are diagnosable. To verify this fact, we can substitute y_1 and y_2 in the system (3.23)–(3.26), then

$$f_1 = \dot{y}_2 - y_1 y_2 \tag{3.28}$$
$$f_2 = \dot{y}_1 - \dot{y}_2 + y_1 y_2 - y_2. \tag{3.29}$$

It can be said that system (3.23)–(3.26) is diagnosable and the faults can be reconstructed from the knowledge of y_1, y_2 and their time derivatives.

Remark 3.2 The diagnosability condition is independent of the observability of a system [8].

3.2 Invariant Observer

We briefly recall here some main definitions [2].

Definition 3.2 Let G be a Lie group with identity e and Σ open set. A transformation group $(\phi_g)_{g \in G}$ on is a smooth map

$$\phi : G \times \Sigma \to \Sigma \tag{3.30}$$
$$(g, \xi) \to \phi_g(\xi) \tag{3.31}$$

such that

$$\phi_g(\xi) = \xi \quad \forall \xi \tag{3.32}$$

$$\phi_{g_2} \circ \phi_{g_1} = \phi_{g_2 g_1}(\xi) \quad \forall g_1, g_2, \xi. \tag{3.33}$$

Consider now the smooth output system.

$$\dot{x} = f(x, u) \tag{3.34}$$

$$y = h(x, u) \tag{3.35}$$

where x belongs to an open subset $X \subset \mathbb{R}^n$, u to an open subset $U \subset \mathbb{R}^m$ and y to an open subset $Y \subset \mathbb{R}^p$, $p \leq n$. We assume the signals $u(t)$, $y(t)$ are known (y is a measured signal, and $u(t)$ is known as the control input).

Consider also the local group of transformations on $X \times U$ defined by $(X, U) = (\phi_g(x), \psi_g(u))$, where ϕ_g and $\psi_g(u)$ are local diffeomorphisms.

Definition 3.3 The system $\dot{x} = f(x, u)$ is *G-invariant* if $f(\phi(x), \psi(u)) = \frac{\partial}{\partial x}$ $(\phi_g(x)) \cdot f(x, u)$ for all g, x, u. The property also reads $\dot{X} = f(X, U)$, i.e., the system is left unchanged by the transformation.

Definition 3.4 The output $y = h(x, u)$ is *G-compatible* if there exist a transformation group $(\rho_g)_{g \in \mathbb{R}}$ on Y such that $h(\phi_g(x), \psi_g(u)) = \rho_g(h(x, u))$ for all g, x, u.

With $(X, Y) = (\phi_g(x), \psi_g(u))$ and $Y = \rho_g(y)$, the definition reads $Y = h(X, U)$.

3.2.1 Characterization of Invariant Observers

Definition 3.5 The observer is *G-invariant* if

$$F(\phi_g(\hat{x}), \psi_g(u), \rho_g(y)) = \frac{\partial}{\partial x}(\phi_g(x)) \cdot F(\hat{x}, u, y), \text{ for all } g, \hat{x}, u, y \tag{3.36}$$

The property also reads $\dot{\hat{X}} = F(\hat{X}, U, Y)$, i.e., the system is left unchanged by transformation. An observer is then a pre-observer such that $\hat{x} \to x(t)$ (possibly only locally).

Remark 3.3 In general, the geometry of the sit is not preserve with the "usual" output error $\hat{y} - y = h(\hat{x}, u) - y$, hence it will not yield an invariant observer. The key idea in order to build an invariant observer is to use an invariant output error instead of the usual output error.

Definition 3.6 The smooth map $(\hat{x}, u, y) \to E(\hat{x}, u, y) \in Y$ is an invariant output error if

(a) The map $y \to E(\hat{x}, u, y)$ is invertible for all $\hat{x}, u$.
(b) $E(\hat{x}, u, h(\hat{x}, u)) = 0$ for all $\hat{x}, u$.

(c) $E(\phi_g(\hat{x}), \psi_g(u), \rho_g(y)) = E(\hat{x}, u, y)$.

The first and second property mean E is an output error, i.e., it is zero if and only if $h(\hat{x}, u) = y$; the third property, which also reads $E(\hat{X}, U, Y) = E(\hat{x}, u, y)$, defines invariance.

Now an application [2] $x \to \gamma(x)$ is obtained by solving for g the so-called normalization equation $\phi_g(x) = c$ for some arbitrary constant c; in other words $\phi_{\gamma(x)}(x) = 0$.

Theorem 3.2 ([2]) *The general invariant observer reads*

$$F(\hat{x}, u, y) = f(\hat{x}, u) + \sum_{i=1}^{n}(L_i(E, I)) \cdot E)\omega_i(\hat{x}) \tag{3.37}$$

where

(a) ω_i, $i = 1, \ldots, n$ is the invariant vector field defined by

$$\omega_i(\hat{x}) = \left[\frac{\partial}{\partial \hat{x}_i} \phi_{\gamma(x)} \right]^{-1} \cdot \frac{\partial}{\partial x_i} \tag{3.38}$$

with $\frac{\partial}{\partial x_i}$ the ith canonical vector field on X
(b) E is the invariant error defined by

$$E(\hat{x}, u, y) = \rho_{\gamma(\hat{x})}(h(\hat{x}, u)) - \rho_{\gamma(\hat{x})}(y) \tag{3.39}$$

(c) I is the (complete) invariant set defined by

$$I(\hat{x}, u) = \psi_{\gamma(x)}(u) \tag{3.40}$$

(d) L_i, $i = 1, \ldots, n$ is a $1 \times p$ matrix with entries possibly depending on E and I, and can be freely chosen.

Finally, the observer also can be written in the form

$$F(\hat{x}, u, y) = f(\hat{x}, u) + W(\hat{x})L(I(\hat{x}, u), E(\hat{x}, u, y))E(\hat{x}, u, y) \tag{3.41}$$

where $W(\hat{x}) = (\omega_1(\hat{x}), \ldots, \omega_n(\hat{x}))$ and L is $n \times p$ matrix whose entries depend on (I, E). The observer can be thought of as a gain-scheduled observer with a $n \times p$ gain matrix $W \cdot L$ multiplied by the non-linear error E.

3.3   Construction of the Observer

3.3.1   Reduced-Order Observer

Let consider system (3.1). The fault vector is unknown and it can be assimilated as a state with uncertain dynamics. Then, in order to estimate it, the state vector is extended to deal with the unknown fault vector. The new extended system is given by

$$\dot{x}(t) = A(x, \bar{u}) \tag{3.42}$$

$$\dot{f} = \Omega(x, \bar{u}) \tag{3.43}$$

$$y(t) = h(x, u) \tag{3.44}$$

where $\Omega(x, \bar{u}) = [\Omega_1(x, \bar{u}), \ldots, \Omega_\mu(x, \bar{u})] : \mathbb{R}^{n+m+\mu} \to \mathbb{R}^\mu$ is an uncertain function. Note that a classic Luenberger observer can not be constructed the term $\Omega(x, \bar{u})$ is unknown. This problem is overcome by using a reduced-order uncertainty observer in order to estimate the fault signal f. Next Lemma describes the construction of a proportional reduced-order observer for the system showed in Eqs. (3.42)–(3.44).

Lemma 3.1 ([3]) *If the following hypotheses are satisfied*

H_1: *$\Omega(x, \bar{u})$ is bounded, i.e. $\|\Omega(x, \bar{u})\| \leq N \in \mathbb{R}^+ \, \forall i, \, 1 \leq i \leq \mu$*
H_2: *$f(t)$ is algebraically observable over $\mathbb{R}\langle u, y \rangle$ then the system*

$$\dot{\hat{f}} = k_i(f_i - \hat{f}_i) \quad 1 \leq i \leq \mu \tag{3.45}$$

is a reduced order observer for system (7), where $\hat{f}_i$ denotes the estimate of fault f_i and $k_i \in \mathbb{R}^+ \forall i = 1, \ldots, \mu$ are positive real coefficients that determine the desired convergence rate of the observer.

Lemma 3.2 ([8]) *If a fault signal f_i, $i \in \{1, \ldots, \mu\}$ of system (1) is algebraic observable and can be written in the following form*

$$f_i = a_i \dot{y} + b_i(u, y) \tag{3.46}$$

where $a_i = [a_{i1}, \ldots, a_{im}] \in \mathbb{R}^m$ is a constant vector and $b_i(u, y)$ is a bounded function, then there exists a function $\gamma_i \in C^1$, such that the reduced order observer (3.45) can be written as the following asymptotically stable system.

$$\dot{\gamma}_i = -k_i \gamma_i + k_i b_i(u, y) - k_i^2 a_i y \tag{3.47}$$

$$\hat{f} = \gamma_i + k_i a_i y \tag{3.48}$$

with $\gamma_i(0) = \gamma_{i0} \in \mathbb{R}$.

3.3.2 Invariant Observer

Consider the nonlinear system with faults given by (3.1), assuming that the fault vector f is algebraically observable over $\mathbb{R}\langle u, y \rangle$, therefore, it satisfies a differential algebraic polynomial.

$$\bar{\psi}(f, y, \dot{y}, \ddot{y}, \ldots, y^{(r)}, u, \dot{u}, \ldots) = 0 \tag{3.49}$$

where r is the maximum order of the output time derivatives.

Now, introducing the following change of coordinates

$$\eta_1 = y \tag{3.50}$$
$$\eta_2 = \dot{y} \tag{3.51}$$
$$\vdots$$
$$\eta_r = y^{(r-1)} \tag{3.52}$$

we obtain the following representation form the coordinates (3.50)–(3.52) which is the so-called Generalized Observability Canonical Form (GOCF).

$$\dot{\eta}_1 = \eta_2 \tag{3.53}$$
$$\dot{\eta}_2 = \eta_3 \tag{3.54}$$
$$\vdots$$
$$\dot{\eta}_r = \Psi(f, \eta_1, \eta_2, \eta_3, \ldots, \eta_r, u, \dot{u}, \ldots) \tag{3.55}$$
$$y = \eta_1 \tag{3.56}$$

where $\Psi(\cdot)$ is considered as an unmodeled dynamics.

Now, consider the system (3.53)–(3.56) and take $r = 2$. The following scaling transformation group

$$\varphi_g \begin{pmatrix} \eta_1 \\ \eta_2 \end{pmatrix} = \begin{pmatrix} g\eta_1 \\ g\eta_2 \end{pmatrix} \tag{3.57}$$
$$\rho_g(y) = (gy) \tag{3.58}$$

assuming that the $\phi(\cdot)$ dynamics is invariant to the change of coordinates $\varphi_g(\eta)$, the transformations (3.57)–(3.58) satisfy Definitions 3.2 and 3.3, which implies that system (3.53)–(3.56) with $r = 2$ is G-invariant by the transformation group φ_g, and the output is G-compatible over ρ_g.

To find the $\gamma(\eta)$ we choose the first component of φ_g for the normalization equation, $g\eta_1 = 1$ and get

$$\gamma(\eta_1) = \left(\frac{1}{\eta_1}\right) \tag{3.59}$$

We get then the invariant error

$$E(\hat{\eta}, u, y) = \rho_{\gamma(\hat{\eta}_1)}(\hat{y}) - \rho_{\gamma(\hat{\eta}_1)}(y) \tag{3.60}$$

$$E = \frac{\hat{\eta}_1 - \eta_1}{\hat{\eta}_1} = \frac{\hat{y} - y}{\hat{y}} \tag{3.61}$$

To build an invariant frame, we apply the following

$$\omega_i(\hat{\eta}) = \left[\frac{\partial}{\partial \hat{\eta}_i} \varphi_{\gamma(\hat{\eta})}(\hat{\eta})\right]^{-1} \cdot \frac{\partial}{\partial \eta_i}, i = 1, \dots, n \tag{3.62}$$

Since $\left[\frac{\partial}{\partial \hat{\eta}_i} \varphi_{\gamma(\hat{\eta})}(\hat{\eta})\right]^{-1} = \frac{\partial}{\partial \hat{\eta}_i} \varphi_{\gamma^{-1}(\hat{\eta})}(\hat{\eta})$ and $\gamma^{-1} = (\hat{\eta})$, then the invariant frame is

$$W(\hat{\eta}) = [\omega_1, \omega_2] \tag{3.63}$$

$$W(\hat{\eta}) = \frac{\partial}{\partial \hat{\eta}_i} \varphi_{\hat{\eta}_1}(\hat{\eta}), 1 \leq i \leq 2 \tag{3.64}$$

$$W(\hat{\eta}) = \frac{\partial}{\partial \hat{\eta}_i} \begin{pmatrix} \hat{\eta}_1^2 \\ \eta_1 \eta_2 \end{pmatrix} \tag{3.65}$$

$$W(\hat{\eta}) = \begin{bmatrix} 2\hat{\eta}_1 & 0 \\ \hat{\eta}_2 & \hat{\eta}_1 \end{bmatrix} \tag{3.66}$$

Finally, we choose the vector $L = [L_1, L_2]^T$ as a positive real vector, yielding the invariant observer as follows

$$\dot{\hat{\eta}}_1 = \eta_2 + (2L_1 \hat{\eta}_1) \left(\frac{\hat{y} - y}{\hat{y}}\right) \tag{3.67}$$

$$\dot{\hat{\eta}}_2 = (L_1 \hat{\eta}_2 + L_2 \hat{\eta}_1) \left(\frac{\hat{y} - y}{\hat{y}}\right) \tag{3.68}$$

$$\hat{y} = \eta_1 \tag{3.69}$$

Then returning to the original coordinates and taking into account (3.49), the fault signal can be estimated from the following relationship.

$$\hat{\bar{\psi}}(\hat{f}, \eta, \dot{\eta}, \hat{\eta}, \ddot{\hat{\eta}}, \dots, \hat{\eta}^{(r)}, u, \dot{u}, \ddot{u}, \dots) = 0. \tag{3.70}$$

3.4 Real-Time Application

3.4.1 Description of the Three-Tank System

The Amira DTS200 system is described in Fig. 3.1. The dynamic with faults is described by the following equations [8] where f_1 and f_2 ($\mu = 2$) are considered in the actuators that control the input flow. They could be originated by an electronic component malfunction, or even by a leakage or an obstruction in the pump pipes.

$$\dot{x}_1 = \frac{1}{A}(u_1 - q_{13} + f_1) \tag{3.71}$$

$$\dot{x}_2 = \frac{1}{A}(u_2 + q_{32} - q_{20} + f_2) \tag{3.72}$$

$$\dot{x}_3 = \frac{1}{A}(q_{13} - q_{32}) \tag{3.73}$$

$$y_1 = x_2 \tag{3.74}$$

$$y_2 = x_3 \tag{3.75}$$

In this system, $u_i = q_i$, $i = 1, 2$ are the manipulable input flows, $x_i = h_i$, $i = 1, 2, 3$ are the levels of each tank, A is the cross section of the tanks, and the terms q_{ij} represent the water flow from tank i to tank j. S is the cross-sectional area of the pipe that interconnects each tank and the unknown parameters a_i, $i = 1, 2, 3$ are the output flow coefficients which are not known, so they are considered as uncertain parameters.

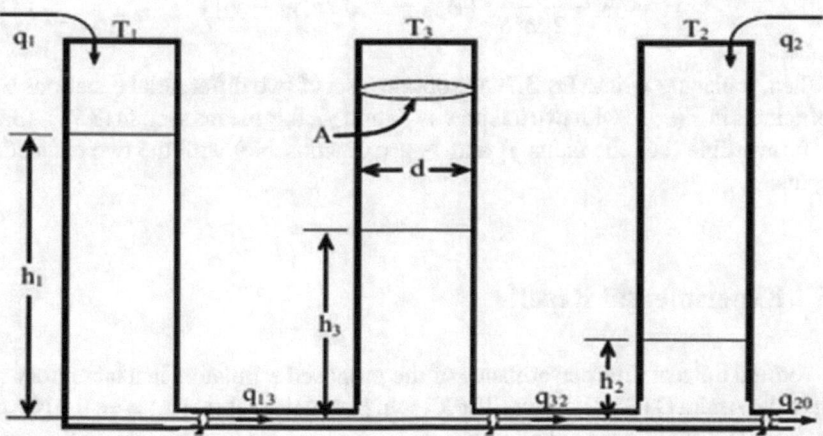

Fig. 3.1 Scheme diagram of the three-tank system

$$q_{13} = a_1 S \sqrt{2g|x_1 - x_3|} \tag{3.76}$$

$$q_{32} = a_3 S \sqrt{2g|x_3 - x_2|} \tag{3.77}$$

$$q_{20} = a_2 S \sqrt{2g|x_2|} \tag{3.78}$$

The system (3.71)–(3.75) has four state regions in which the corresponding model is differentiable, here $x_1 \geq x_2 \geq x_3$ is the region considered for the analysis.

According to Theorem 3.1 it is necessary measure two or more outputs; so, this scenery can happen only in the following cases:

- Case 1, where $p = 3$ (All measurable states).
- Case 2, where $p = 2$ (h_1 not measurable).
- Case 3, where $p = 2$ (h_2 not measurable).
- Case 4, where $p = 2$ (h_3 not measurable).

Here is considered Case 2, where $h_1 = x_1$ is not measurable.

3.4.2   Diagnosability Analysis

According to Theorem 3.1 we need two or more measurable outputs in order to system (3.1) be classified as left invertible. The case when x_1 is not a measurable signal is considered. Taking into account Eqs. (3.71)–(3.75) we have the follows

$$f_1 = A\dot{x}_1 + a_1 S \sqrt{2g|x_1 - y_2|} - u_1 \tag{3.79}$$

$$f_2 = A\dot{y}_2 - a_3 S \sqrt{2g|y_3 - y_2|} + a_2 S \sqrt{2g|y_2|} - u_2 \tag{3.80}$$

$$x_1 = y_3 - \frac{1}{2ga_1^2 S^2} \left(A\dot{y}_3 - a_3 \sqrt{2g|y_3 - y_2|} \right)^2 \tag{3.81}$$

Then, replacing x_1 into Eq. 3.79 we obtain a set of two differential equations with coefficients in $\mathbb{R}\langle u, y \rangle$ with two unknowns f_1 and f_2, this means system (3.71)–(3.75) is left invertible (i.e., the faults f_1 and f_2 are diagnosable) with the two considered outputs.

3.5   Experimental Results

We verified the real time performance of the proposed estimators in a laboratory setting of the Amira DTS200 system. The known parameter values are: $A = 0.0149$ m^2, $S = 5 \times 10^{-5}$ m^2 and the unknown parameters a_1, a_2 and a_3. The sample time in all the experiments was 0.001 s.

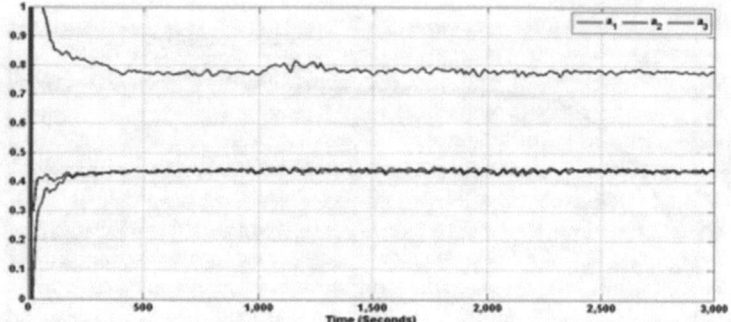

Fig. 3.2 Parameter identification

Before starting the construction of the faults, it is necessary to estimate the flow parameters a_1, a_2 and a_3. Once these parameters have been obtained, the system is monitored, followed by the estimation of the fault signals.

3.5.1 Identification

With no presence of fault, the unknown parameters a_1, a_2 and a_3 were estimate meanwhile the values for the input flow were: $q_1 = 0.00002 \frac{m^3}{s}$ and $q_2 = 0.000015 \frac{m^3}{s}$, along 3000 s in these conditions the evolution of the estimated values for the unknown coefficients is shown in Fig. 3.2.

At the end of the estimation process the estimated values for the flow parameters were obtained:

$$a_1 = 0.4385 \tag{3.82}$$
$$a_2 = 0.7774 \tag{3.83}$$
$$a_3 = 0.4435. \tag{3.84}$$

3.5.2 Fault Estimation Results

In all the experiments in this subsection the input flows were maintained constant as $u_1 = 0.00002 \frac{m^3}{s}$ and for 3000 s, also two faults are artificial generated through the following expressions

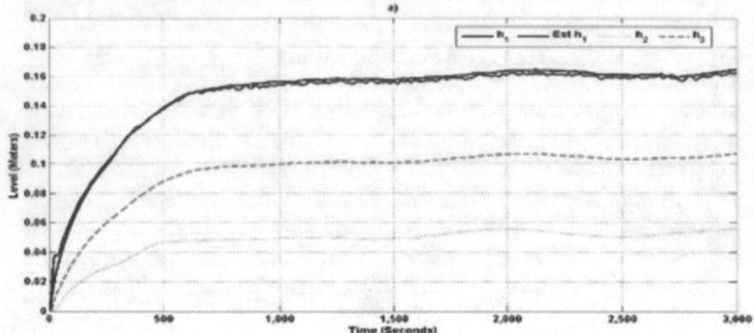

Fig. 3.3 Estimation of the unmeasured state x_1 using the reduced-order observer

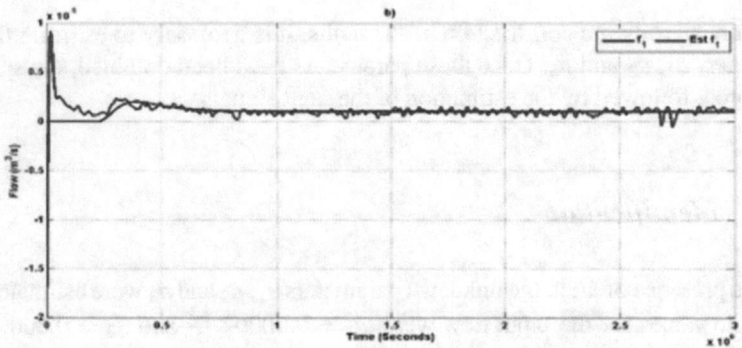

Fig. 3.4 Estimation of the unknown signal f_1 using the reduced-order observer

$$f_1 = 1 \times 10^{-5}[1 + \sin(0.2t \exp^{-0.01t})U(t - 220)] \tag{3.85}$$
$$f_2 = 1 \times 10^{-5}[1 + \sin(0.05t \exp^{-0.001t})U(t - 300)] \tag{3.86}$$

where, $U(\cdot)$ is the step function.

For the two schemes of observers, it is evaluated the case when $x_2 = y_2$ and $x_3 = y_3$ are measurable outputs, because of this, it is necessary to estimate the unknown state x_1. Figures 3.3, 3.4 and 3.5 show the resulting estimations achieved with the reduced-order observer. A low-pass filter was necessary in order to reduce the effect of the measurement noise, we chose a second order Butterworth filter whose transfer function is given by $G_f(s) = \frac{1}{32s^2+8s+1}$. The gain values chosen for the state and fault observers were $k_{x_1} = 0.3$ to estimate x_1, $k_{f_1} = 1.85$ and $k_{f_2} = 22$ to estimate f_1 and f_2.

In the same way, the estimation results using the invariant observer are presented in Figs. 3.6, 3.7 and 3.8. The gains values for each observer were chosen like $L_{x_1} = [5.5.3.5]^T$ to estimate x_1, $L_{f_1} = [1.5, 0.5]^T$ and $L_{f_2} = [3.5, 2.5]^T$ to estimate the fault signals f_1 and f_2, respectively.

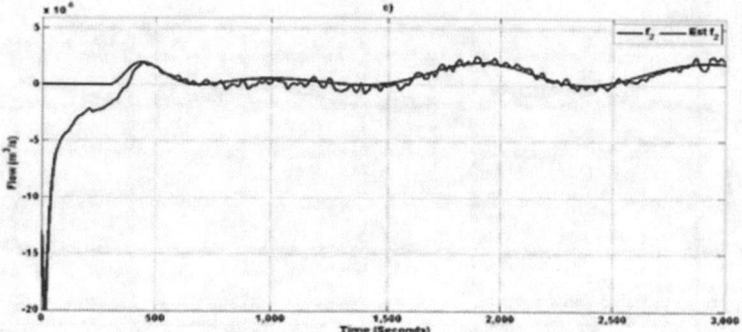

Fig. 3.5 Estimation of the unknown signal f_2 using the reduced-order observer

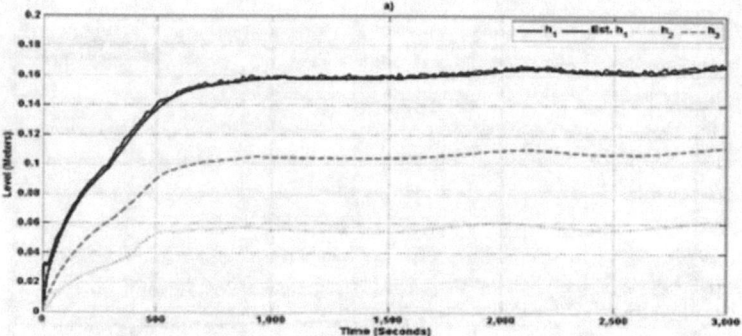

Fig. 3.6 Estimation of the unmeasured state x_1 using the invariant observer

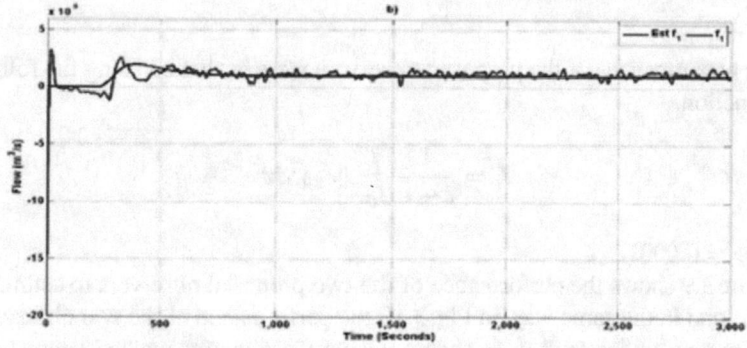

Fig. 3.7 Estimation of the unknown signal f_1 using the invariant observer

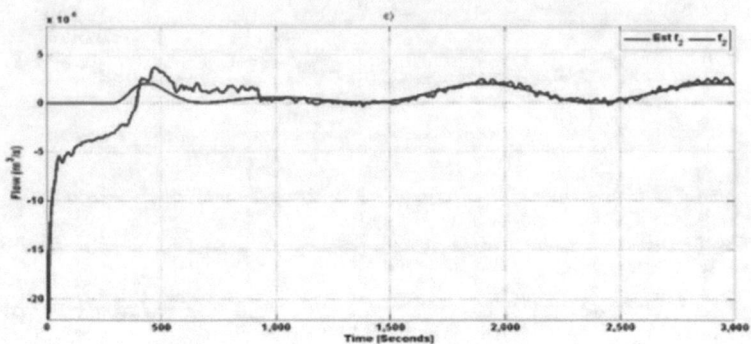

Fig. 3.8 Estimation of the unknown signal f_2 using the invariant observer

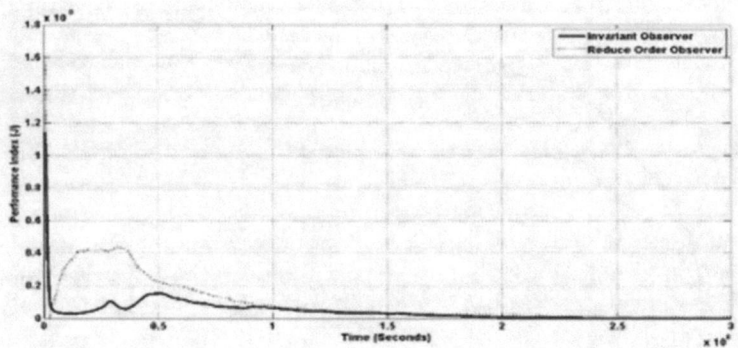

Fig. 3.9 Performance evaluation of observers for the estimation error of the fault f_1

The performances of the proposed observers were evaluated using the following cost function.

$$J_t = \frac{1}{t - \varepsilon} \int_0^t \|e_k\|^2 d\tau \tag{3.87}$$

where $\varepsilon = 0.0001$.

Figure 3.9 shows the performance of the two proposed observers to estimate the fault f_1, and in the same way, in Fig. 3.10 the performance of the two observers for the estimation of the fault f_2 is shown. Using these results we can guarantee that the invariant observer offers a better performance to estimate fault signals compared with the reduced-order observer.

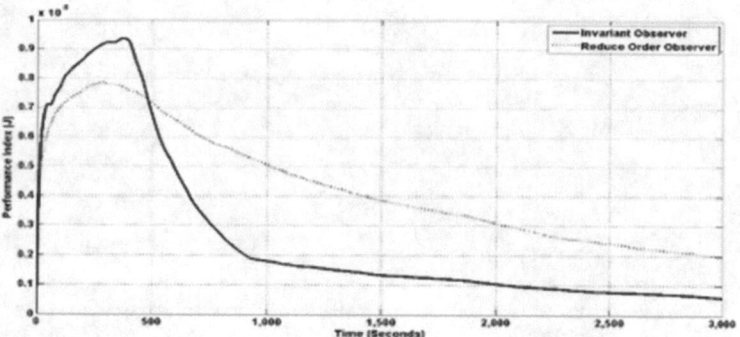

Fig. 3.10 Performance evaluation of observers for the estimation error of the fault f_2

3.6 Conclusions

We have tackled with the diagnosis problem in non-linear systems using the condition of left invertibility through the concept of differential output rank to guarantee the diagnosability of the system. The theoretical and simulation results were tested in a real-time implementation (the three-tank system Amira DTS200) where the experiments showed a better performance when the diagnosis of faults and states via invariant observer is performed. The fluctuation of the parameter makes hard to implement the observers, however, the results show that both observers can reconstruct the fault despite of variations of such parameters.

References

1. Alcorta-García, E., Frank, P.: Deterministic nonlinear observer-based approaches to fault diagnosis: a survey. Control Eng. Pract. **5**(5), 663–670 (1997)
2. Bonnabel, S., Martin, P., Rouchon, P.: Invariant Observers (2007). arxiv.math.OC/0612193v2. Submitted to IEEE Transactions on Automatic Control
3. Cruz-Victoria, J.C., Martínez-Guerra, R., Rincón-Pasaye, J.J.: On nonlinear systems diagnosis using differential and algebraic methods. J. Frankl Inst. **345**(2), 102–118 (2008)
4. Fliess, M.: A note on invertibility of nonlinear input output differential systems. Syst. Control Lett. **8**(2), 147–151 (1986)
5. Kolchin, E.R.: Differential Algebra and Algebraic Groups. Academic, New York (1973)
6. Martínez-Guerra, R., Diop, S.: Diagnosis of nonlinear systems using an unknown-input observer: an algebraic and differential approach. IEE Proc.-Control Theory Appl. **151**(1), 130–135 (2004)
7. Martínez-Guerra, R., Mata-Machuca, J.L., Rincón-Pasaye, J.J.: Fault diagnosis viewed as a left invertibility problem. ISA Trans. **52**(5), 652–661 (2013)
8. Martínez-Guerra, R., Mata-Machuca, J.L.: Fault Detection and Diagnosis in Nonlinear Systems: A Differential and Algebraic Viewpoint. Springer, Cham (2014)
9. Willsky, A.: A survey of design methods in observer-based fault detection systems. Automatica **1**(2), 601–611 (1976)

Chapter 4
Multi-Fault-Tolerant Control in Integer-Order Systems

Controls with fault tolerance have a key role in many applications in automation and engineering [4, 13, 14, 20]. There are many different approaches to achieve such control, for instance the survey [19] provides a basic literature review covering most areas of fault-tolerant control (FTC). The book [3] presents a model-based approach for FTC. Particularly in the field of nonlinear systems, encouraging results were recently obtained applying algebraic techniques [7, 11–14]. This paper proposes to solve this problem the construction of a fault-tolerant dynamical controller capable of linearizing the tracking errors dynamics; the controller is obtained transforming the system into a multi-input multi-output generalized observability canonical form (MMGOCF) represented as a chain of integrators.

For MIMO systems with faults, the dynamical controller is dependent on each of them, forcing the implementation of a fault diagnosis (FD) method. Then, if the system has multiple faults (additive and multiplicative), the FD method has to be able to reconstruct each fault simultaneously and online.

FD has been a research area for a long time. There are numerous works that study this problem, as [1, 22], involving residual generation, disturbance decoupling, and adaptive approaches. For linear systems, the geometric approach has also been applied [18], where the concept of unobservability subspaces is used together with residual generators. There are also several papers that study fault detection and diagnosis in nonlinear systems [9, 10, 23]. For example, in [10] a Gaussian particle filter is used for estimation purpose; as another example, an adaptive estimation algorithm for recursive estimation of parameters related to faults is the way that the paper [23] deals with the problem. Also, the geometric approach for failure detection and isolation via residual generation has been extended to nonlinear systems [6], where the distribution tools are considered as the main ingredients of the unobservability subspaces. Furthermore, the fault diagnosis under the data-driven framework has received great attention recently [24, 25].

© The Author(s), under exclusive license to Springer Nature Switzerland AG 2021
R. Martínez-Guerra et al., *Fault-tolerant Control and Diagnosis for Integer and Fractional-order Systems*, Studies in Systems, Decision and Control 328,
https://doi.org/10.1007/978-3-030-62094-3_4

The solution proposed to the problem of diagnosis of additive and multiplicative faults is performed using the differential algebraic approach [15, 16]. In the book [16] the fault detection and diagnosis problem in nonlinear systems is presented using differential and algebraic tools. This framework uses the reduced-order observer (model-free type) to achieve an effective estimation of the faults, while giving the possibility to estimate online and simultaneously several faults [17, 21].

On the other hand, in order to implement tracking on the MIMO system, it is necessary to use also an observer. A high-gain observer (HGO) [8] is employed for estimating the tracking error dynamics and assure its stability. Then, it is proven that the overall system is ultimate uniformly bounded in the presence of measurement noise [5].

Furthermore, some real-time results are presented using the Amira DTS200 system [2]. The Amira system provides an opportunity to introduce multiple faults in sensors and actuators, making it a very versatile system (benchmark); this is why it has been widely used for experimental studies on FD and FTC [7, 11, 12, 17]. In particular, in [7] the fault diagnosis is realized via algebraic estimation of derivatives, that yields estimates of the residuals (fault indicators), whereas in this paper the faults are not only detected but also diagnosed as they are reconstructed via an observer. Another difference lies in the control laws. While in [7] is used a nonlinear extension of a classic proportional-integral (PI) controller, that is independent of the faults, in the current paper is proposed a dynamical controller. This dynamical controller seeks to stabilize the tracking error system and depends on simultaneous fault diagnosis (fault-tolerant dynamical controller), which leads to the elimination of the effects of the faults in the system. As far as we know, this technique has not been employed in literature.

4.1   Fault Diagnosis

The dynamical controller proposed for fault-tolerant tracking is dependent on the multiple faults that appear in the system. Thus, the need for a FD is a consequence of the proposed method, and a diagnosis capable to reconstruct each fault is required.

Firstly, consider a nonlinear systems with faults, described by:

$$\dot{x} = g(x, u, f) \tag{4.1}$$
$$y = h(x, u)$$

where $x \in \mathbb{R}^n$ is the state vector, $u \in \mathbb{R}^m$ the control input vector, $f \in \mathbb{R}^r$ is an unknown input vector (fault vector), $y \in \mathbb{R}^p$ is the output vector, and g and h are assumed to be analytic functions.

As mentioned above, the estimation of the faults that appear in the system is needed. To achieve this in the differential algebraic framework, the system has to satisfy the diagnosability property.

Definition 4.1 An element f in the differential field $k \langle u, y \rangle$ is said to be algebraically observable with respect to u and y if it satisfies a differential equation with coefficients over $k \langle u, y \rangle$ (k is a constant field, u and y are differential quantities).

Definition 4.2 A system of the form (4.1) is said to be diagnosable if it is possible to estimate the fault f from the system equations and the time histories of the data u and y. This is, the system is diagnosable if f is algebraically observable with respect to u and y.

The notion of algebraic observability is to express the faults as polynomial equations dependent on the inputs and outputs of the system and finitely many time derivatives, with coefficients in k:

$$f_{\bar{\imath}} = P_{\bar{\imath}}(u, \dot{u}, \ldots, y, \dot{y}, \ldots)$$

In order to illustrate Definition 4.2, two examples are presented.

Example 4.1 Let us consider the following nonlinear system.

$$\begin{aligned}
\dot{x}_1 &= x_1 x_2 + f_1 + u \\
\dot{x}_2 &= x_1 \\
y_1 &= x_1 + f_2 \\
y_2 &= x_2
\end{aligned} \tag{4.2}$$

Since f_1 and f_2 satisfy Definition 4.2

$$\begin{aligned}
f_1 &= \ddot{y}_2 - y_2 \dot{y}_2 - u \\
f_2 &= y_1 - \dot{y}_2
\end{aligned}$$

the system (4.2) is diagnosable and the faults can be reconstructed from the knowledge of u, y and their time derivatives.

Example 4.2 The system

$$\begin{aligned}
\dot{x}_1 &= (x_1 + x_2)(u + f) \\
\dot{x}_2 &= u \\
y &= x_1 + x_2
\end{aligned} \tag{4.3}$$

is a diagnosable system since

$$f = \frac{\dot{y} - u}{y} - u.$$

It is clear that systems with additive and multiplicative faults can be addressed using the proposed algebraic approach.

Moreover, the unknown fault vector $f = (f_1, \ldots, f_r)$ can be seen as a state with an uncertain dynamics $\Omega (x, u, f) : \mathbb{R}^{n+m+r} \to \mathbb{R}^r$. Then, in order to estimate it, the state vector is extended (immersion [16]).

As it can be seen a classic Luenberger observer, which needs full knowledge of the system dynamics, can not be constructed because the term $\Omega (x, u, f)$ is unknown. However this problem can be solved using a reduced-order observer (ROO), because it can be implemented from the algebraic observability property of the faults, and is asymptotically stable. Next lemma describes the construction of a proportional ROO for (4.1).

Lemma 4.1 ([16]) *If the following hypotheses are satisfied:*
  Assumption 1 $\left| \Omega_{\bar{l}} (x, u, f) \right| \leq N_{\bar{l}} \in \mathbb{R}^+$ $\quad \forall \bar{l} = 1, \ldots, r.$
  Assumption 2 $f (t)$ *is algebraically observable over* $\mathbb{R} \langle u, y \rangle.$
  Then the system

$$\dot{\hat{f}}_{\bar{l}} = k_{\bar{l}}(f_{\bar{l}} - \hat{f}_{\bar{l}}) \qquad 1 \leq \bar{l} \leq r \qquad (4.4)$$

is an asymptotic reduced-order observer for system (4.1), where $\hat{f}_{\bar{l}}$ denotes the estimate of fault $f_{\bar{l}}$ and $k_{\bar{l}} \in \mathbb{R}^+$, $\forall \bar{l} = 1, \ldots, r$ are positive coefficients that determine the desired convergence rate of the observer.

Sometimes the output derivatives appear in the algebraic equation of the fault, then it is necessary to use an auxiliary variable to approximate them as described in the next lemma.

Lemma 4.2 ([16]) *If a fault signal $f_{\bar{l}}$, $1 \leq \bar{l} \leq r$ of system (4.1) is algebraically observable and can be written in the following form:*

$$f_{\bar{l}} = a_{\bar{l}} \dot{y} + b_{\bar{l}} (u, y) \qquad (4.5)$$

where $a_{\bar{l}} = [a_1, \ldots, a_m] \in \mathbb{R}^m$ is a constant vector and $b_{\bar{l}}(u, y)$ is a bounded function, then there exists a function $\gamma_{\bar{l}} \in C^1$, such that the reduced-order observer (4.4) can be written as the following asymptotically stable system:

$$\dot{\gamma}_{\bar{l}} = -k_{\bar{l}} \gamma_{\bar{l}} + k_{\bar{l}} b_{\bar{l}} (x, u) - k_{\bar{l}}^2 a_{\bar{l}} y \qquad \gamma_{\bar{l}} (0) = \gamma_{\bar{l}0} \in \mathbb{R} \qquad (4.6)$$
$$\hat{f}_{\bar{l}} = \gamma_{\bar{l}} + k_{\bar{l}} a_{\bar{l}} y.$$

Remark 4.1 This methodology is recursive; we can introduce as many virtual variables as needed.

Remark 4.2 The ROO also serves as an estimator of derivatives. If there are output time derivatives of order 2 or higher, consider the time derivative to be estimated $\eta = \dot{y}$. According to (4.4), we propose the observer structure

$$\dot{\hat{\eta}} = k_\eta(\eta - \hat{\eta}) \qquad (4.7)$$

introducing the change of variable $\gamma = \hat{\eta} - k_\eta y$, and from (4.7), we have

$$\dot{\gamma} = -k_\eta \hat{\eta}$$
$$= -k_\eta \gamma - k_\eta^2 y \qquad (4.8)$$

which constitutes with $\hat{\eta}$ an asymptotic estimator for $\eta = \dot{y}$.

Now, considering the ROO dynamics, the following variables are defined:

$$\hat{f}_{i\bar{l}} = \hat{f}_{\bar{l}}^{(i-1)} \qquad i = 1, \ldots, \mu_{\bar{l}} \qquad (4.9)$$

then the fault estimation subsystems are written as:

$$\dot{\mathbf{f}}_{\bar{l}} = E\hat{\mathbf{f}}_{\bar{l}} + \omega_{\bar{l}}(u, y, f) \qquad 1 \leq \bar{l} \leq r \qquad (4.10)$$

where

$$\hat{\mathbf{f}}_{\bar{l}} = \left(\hat{f}_{1\bar{l}}, \ldots, \hat{f}_{\mu_{\bar{l}}, \bar{l}} \right) \qquad \omega_{\bar{l}}(u, y, f) = \begin{pmatrix} 0 \\ \vdots \\ 0 \\ k_{\bar{l}}(f_{\bar{l}}^{(\mu_{\bar{l}}-1)} - \hat{f}_{\mu_{\bar{l}}, \bar{l}}) \end{pmatrix}$$

Remark 4.3 The ROO is able to reconstruct the states, the faults and their time derivatives, provided that they satisfy the conditions of Lemma 4.1. Given that this procedure can be done online, this is useful to perform real-time applications.

4.2 Fault-Tolerant Control

A nonlinear system described by (4.1) can be represented by the following multi-input multi-output generalized observability canonical form (MMGOCF) due to the differential primitive element:

$$\dot{\eta}_{ij} = \eta_{i+1, j} \qquad 1 \leq i \leq n - 1 \qquad (4.11)$$
$$\dot{\eta}_{nj} = -L_j(\eta_1, \ldots, \eta_p, u, \ldots, u^{(\gamma)}, f, \ldots, f^{(\mu)})$$
$$y_j = \eta_{1j}$$

where L_j is a C^1 real-valued function, $\eta_j = (\eta_{1j}, \ldots, \eta_{nj}) \in \mathbb{R}^n$, $y \in \mathbb{R}^p$, $u \in \mathbb{R}^m$, $f \in \mathbb{R}^r$, and some integers $\gamma, \mu \geq 0$. This MMGOCF is constituted of p subsystems, one for each output y_j, $1 \leq j \leq p$.

Let $y_R \in \mathbb{R}^p$ be a reference output vector with C^n function elements. The output tracking problem with FT consists in finding a dynamical controller that depends on the reference output vector y_R and its time derivatives $y_R^{(i)}$, the state variables η_{ij} of the canonical system, and, as previously mentioned, an estimation $\hat{f}$ of the fault vector and its time derivatives, such that the controller locally forces y to converge towards y_R.

Define the output tracking error as:

$$e_{1j} = y_j - y_{Rj} \qquad\qquad 1 \leq j \leq p \tag{4.12}$$

Given that η_{ij} is equal to the $(i-1)$th time derivative of y_j, that is $\eta_{ij} = y_j^{(i-1)}$, for $1 \leq i \leq n$ and $1 \leq j \leq p$, the error variables are rewritten as:

$$e_{1j} = \eta_{1j} - y_{Rj} \qquad\qquad 1 \leq j \leq p \tag{4.13}$$

The p output errors define the following MMGOCF:

$$e_j^{(i)} = \eta_{i+1,j} - y_{Rj}^{(i)} \qquad\qquad 1 \leq i \leq n-1 \tag{4.14}$$
$$e_j^{(n)} = \dot{\eta}_{nj} - y_{Rj}^{(n)} = -L_j(\eta_1, \ldots, \eta_p, u, \ldots, u^{(\gamma)}, \hat{f}, \ldots, \hat{f}^{(\mu)}) - y_{Rj}^{(n)}$$

with $e_{ij} = e_j^{(i-1)}, 1 \leq i \leq n$. Now, a linear time-invariant (LTI) dynamics is imposed for the tracking error:

$$e_j^{(n)} + \sum_{i=0}^{n-1} a_{i+1,j} e_j^{(i)} = 0 \tag{4.15}$$

and from system (4.14), (4.15) is rewritten as:

$$\dot{\eta}_{nj} - y_{Rj}^{(n)} + \sum_{i=1}^{n} a_{ij} \left[\eta_{ij} - y_{Rj}^{(i-1)} \right] = 0 \tag{4.16}$$

that is:

$$-L_j(\eta_1, \ldots, \eta_p, u, \ldots, u^{(\gamma)}, \hat{f}, \ldots, \hat{f}^{(\mu)}) - y_{Rj}^{(n)} = -\sum_{i=1}^{n} a_{ij} \left[\eta_{ij} - y_{Rj}^{(i-1)} \right] \tag{4.17}$$

We can obtain a chain of integrators of the error as follows:

$$\dot{e}_{ij} = e_{i+1,j} \qquad\qquad 1 \leq i \leq n-1$$
$$\dot{e}_{nj} = -\sum_{i=1}^{n} a_{ij} e_{ij} \tag{4.18}$$

or in a compact form:

$$\dot{\mathbf{e}}_j = F_j \mathbf{e}_j \qquad (4.19)$$

and

$$- L_j(\mathbf{e}_1 + \mathbf{y}_{R1}, \dots, \mathbf{e}_p + \mathbf{y}_{Rp}, u, \dots, u^{(\gamma)}, \hat{f}, \dots, \hat{f}^{(\mu)}) - y_{Rj}^{(n)} = - \sum_{i=1}^{n} a_{ij} e_{ij} \qquad (4.20)$$

where $\mathbf{e}_j = (e_{1j}, \dots, e_{nj})$, $\mathbf{y}_{\mathbf{R}j} = (y_{Rj}, \dot{y}_{Rj}, \dots, y_{Rj}^{(n-1)})$, and

$$F_j = \begin{pmatrix} 0 & 1 & \cdots & 0 \\ \vdots & \vdots & \cdots & \vdots \\ 0 & \vdots & \cdots & 1 \\ -a_{1j} & -a_{2j} & \cdots & -a_{nj} \end{pmatrix}$$

The origin $\mathbf{e}_j = 0$ is an equilibrium point for system (4.19) if F_j is Hurwitz.

Furthermore, the controller depends also on the tracking errors. To estimate them, an observer is used. Firstly, system (4.19) is rewritten as:

$$\dot{\mathbf{e}}_j = E \mathbf{e}_j + \varphi_j \left(\mathbf{e}, \mathbf{y}_R, y_{Rj}^{(n)}, \mathbf{u}, \hat{\mathbf{f}} \right) \qquad (4.21)$$

where $\mathbf{e} = (\mathbf{e}_1, \dots, \mathbf{e}_p)$, $\mathbf{y}_R = (\mathbf{y}_{R1}, \dots, \mathbf{y}_{Rp})$, $\mathbf{u} = (u, \dot{u}, \dots, u^{(\gamma)})$, $\hat{\mathbf{f}} = (\hat{f}, \dot{\hat{f}}, \dots, \hat{f}^{(\mu)})$, the elements of E are given by:

$$E_{ks} = \begin{cases} 1 & \text{if } k = s - 1 \\ 0 & \text{otherwise} \end{cases}$$

and

$$\varphi_j \left(\mathbf{e}, \mathbf{y}_R, y_{Rj}^{(n)}, \mathbf{u}, \hat{\mathbf{f}} \right) = \begin{pmatrix} 0 \\ \vdots \\ 0 \\ -L_j(\mathbf{e}_1 + \mathbf{y}_{R1}, \dots, \mathbf{e}_p + \mathbf{y}_{Rp}, u, \dots, u^{(\gamma)}, \hat{f}, \dots, \hat{f}^{(\mu)}) - y_{Rj}^{(n)} \end{pmatrix}$$

Then, the estimation $\hat{\mathbf{e}}_j$ is obtained by the following HGO [8]:

$$\dot{\hat{\mathbf{e}}}_j = E \hat{\mathbf{e}}_j + \varphi_j \left(\hat{\mathbf{e}}, \mathbf{y}_R, y_{Rj}^{(n)}, \mathbf{u}, \hat{\mathbf{f}} \right) - S_{\infty}^{-1} C^T C (\hat{\mathbf{e}}_j - \mathbf{e}_j) \qquad (4.22)$$

where S_{∞} is the solution to the equation:

$$S_\infty \left(E + \frac{\theta}{2} I \right) + \left(E^T + \frac{\theta}{2} I \right) S_\infty = C^T C \tag{4.23}$$

with $\theta > 0$ and $C = \begin{pmatrix} 1 & 0 & \dots & 0 \end{pmatrix}$. The coefficients of S_∞ are given by:

$$(S_\infty)_{ks} = \frac{\alpha_{ks}}{\theta^{k+s-1}}$$

where α_{ks} is a symmetric positive definite matrix independent of θ.

Furthermore, let $\hat{u}_l$ be the solution to

$$-L_j(\hat{\mathbf{e}}, \mathbf{y}_R, \hat{\mathbf{u}}, \hat{u}_l^{(\gamma)}, \hat{\mathbf{f}}) - y_{Rj}^{(n)} = -\sum_{i=1}^{n} a_{ij} \hat{e}_{ij} \tag{4.24}$$

where $\hat{u}_l^{(\gamma)}$ is the highest order derivative of the given input found in the equation. Thus, the dynamical equation for controller $\hat{u}_l$ is:

$$\hat{u}_l^{(\gamma)} = K_l \left(\hat{\mathbf{e}}, \mathbf{y}_R, y_{Rj}^{(n)}, \hat{\mathbf{u}}, \hat{\mathbf{f}} \right) \tag{4.25}$$

These controllers yield tracking in the original MIMO system, with fault tolerance (eliminates the effects of the faults). So, Eq. (4.22) is rewritten as:

$$\dot{\hat{\mathbf{e}}}_j = E\hat{\mathbf{e}}_j + \hat{\varphi}_j \left(\hat{\mathbf{e}}, \mathbf{y}_R, y_{Rj}^{(n)}, \hat{\mathbf{u}}, \hat{\mathbf{f}} \right) - S_\infty^{-1} C^T C (\hat{\mathbf{e}}_j - \mathbf{e}_j) \tag{4.26}$$

with

$$\hat{\varphi}_j \left(\hat{\mathbf{e}}, \mathbf{y}_R, y_{Rj}^{(n)}, \hat{\mathbf{u}}, \hat{\mathbf{f}} \right) = \begin{pmatrix} 0 \\ \vdots \\ 0 \\ -L_j(\hat{\mathbf{e}}_1 + \mathbf{y}_{R1}, \dots, \hat{\mathbf{e}}_p + \mathbf{y}_{Rp}, \hat{u}, \dots, \hat{u}^{(\gamma)}, \hat{f}, \dots, \hat{f}^{(\mu)}) - y_{Rj}^{(n)} \end{pmatrix}$$

Moreover, define the observation error as $\boldsymbol{\varepsilon}_j = \hat{\mathbf{e}}_j - \mathbf{e}_j$, and the following dynamics is obtained from Eqs. (4.21) and (4.26):

$$\dot{\boldsymbol{\varepsilon}}_j = \left(E - S_\infty^{-1} C^T C \right) \boldsymbol{\varepsilon}_j + \Phi_j(\boldsymbol{\varepsilon}, \hat{\mathbf{e}}) \tag{4.27}$$

where

$$\Phi_j(\boldsymbol{\varepsilon}, \hat{\mathbf{e}}) = \hat{\varphi}_j \left(\hat{\mathbf{e}}, \mathbf{y}_R, y_{Rj}^{(n)}, \hat{\mathbf{u}}, \hat{\mathbf{f}} \right) - \varphi_j \left(\underbrace{\hat{\mathbf{e}} - \boldsymbol{\varepsilon}}_{e}, \mathbf{y}_R, y_{Rj}^{(n)} \hat{\mathbf{u}}, \hat{\mathbf{f}} \right)$$

Finally, if the following variables are defined:

$$\hat{u}_{il} = \hat{u}_l^{(i-1)} \qquad\qquad i = 1, \ldots, \gamma_l \qquad\qquad (4.28)$$

then the dynamical controller subsystems are written as follows:

$$\dot{\hat{\mathbf{u}}}_l = E\hat{\mathbf{u}}_l + \kappa_l\left(\hat{\mathbf{e}}, \mathbf{y}_R, y_{Rj}^{(n)}, \hat{\mathbf{u}}, \hat{\mathbf{f}}\right) \qquad\qquad 1 \leq l \leq m \qquad\qquad (4.29)$$

where $\hat{\mathbf{u}}_l = \left(\hat{u}_{1l}, \ldots, \hat{u}_{\gamma_l,l}\right)$ and

$$\kappa_l\left(\hat{\mathbf{e}}, \mathbf{y}_R, y_{Rj}^{(n)}, \hat{\mathbf{u}}, \hat{\mathbf{f}}\right) = \begin{pmatrix} 0 \\ \vdots \\ 0 \\ K_l\left(\hat{\mathbf{e}}, \mathbf{y}_R, y_{Rj}^{(n)}, \hat{\mathbf{u}}, \hat{\mathbf{f}}\right) \end{pmatrix}.$$

Remark 4.4 As it can be seen, the overall FD and FTC system consists on a ROO and a HGO, which represents a hybrid observation system. From this closed-loop system, a separation principle can be determined.

4.3 Stability Analysis of the Closed-Loop System

The closed-loop dynamics is given by:

$$\dot{\hat{\mathbf{e}}}_j = E\hat{\mathbf{e}}_j + \varphi_j\left(\hat{\mathbf{e}}, \mathbf{y}_R, y_{Rj}^{(n)}, \hat{\mathbf{u}}, \hat{\mathbf{f}}\right) - S_\infty^{-1}C^T C(\hat{\mathbf{e}}_j - \mathbf{e}_j)$$
$$\dot{\boldsymbol{\varepsilon}}_j = \left(E - S_\infty^{-1}C^T C\right)\boldsymbol{\varepsilon}_j + \Phi_j(\boldsymbol{\varepsilon}, \hat{\mathbf{e}})$$
$$\dot{\hat{\mathbf{u}}}_l = E\hat{\mathbf{u}}_l + \kappa_l\left(\hat{\mathbf{e}}, \mathbf{y}_R, y_{Rj}^{(n)}, \hat{\mathbf{u}}, \hat{\mathbf{f}}\right) \qquad\qquad (4.30)$$
$$\dot{\hat{\mathbf{f}}}_{\bar{l}} = E\hat{\mathbf{f}}_{\bar{l}} + \omega_{\bar{l}}(u, y, f)$$

for $1 \leq j \leq p$, $1 \leq l \leq m$ and $1 \leq \bar{l} \leq r$. Developing the equations for $\hat{\mathbf{e}}_j$, $\hat{\mathbf{u}}_l$ and $\hat{\mathbf{f}}_{\bar{l}}$, the following chain of integrators is obtained:

$$\dot{\hat{e}}_{ij} = \hat{e}_{i+1,j} - \psi_i\left(\theta_j\right)(\hat{e}_j - e_j) \qquad\qquad 1 \leq i \leq n-1$$
$$\dot{\hat{e}}_{nj} = -L_j(\hat{\mathbf{e}}_1 + \mathbf{y}_{R1}, \ldots, \hat{\mathbf{e}}_p + \mathbf{y}_{Rp}, \hat{u}, \ldots, \hat{u}^{(\gamma)}, \hat{f}, \ldots, \hat{f}^{(\mu)}) - y_{Rj}^{(n)} - \theta_j^n \qquad 1 \leq j \leq p$$
$$\dot{\hat{u}}_{il} = \hat{u}_{i+1,l} \qquad\qquad 1 \leq i \leq \gamma_l - 1$$
$$\dot{\hat{u}}_{\gamma_l,l} = K_l\left(\hat{\mathbf{e}}, \mathbf{y}_R, y_{Rj}^{(n)}, \hat{\mathbf{u}}, \hat{\mathbf{f}}\right) \qquad\qquad 1 \leq l \leq m$$
$$\dot{\hat{f}}_{i\bar{l}} = \hat{f}_{i+1,\bar{l}} \qquad\qquad 1 \leq i \leq \mu_{\bar{l}} - 1$$

$$\dot{\hat{f}}_{\mu_{\bar{l}},\bar{l}} = k_{\bar{l}}(f_{\bar{l}}^{(\mu_{\bar{l}}-1)} - \hat{f}_{\mu_{\bar{l}},\bar{l}})\qquad 1 \le \bar{l} \le r$$

where $\psi_i\left(\theta_j\right)$ is a function obtained from S_∞^{-1}.

In this chain of integrators, the dynamics of the controllers and the fault estimations can be appreciated. As it can be seen, the variables obtained from these dynamics take part explicitly in the tracking error dynamics, leading to the solution of the multi-output tracking problem.

Now the main result of this chapter is stated.

Theorem 4.1 *Let system (4.1) be described in the MMGOCF (4.11) composed of p subsystems. The observation dynamics corresponding to subsystem j are $\hat{\mathbf{e}}_j$ and $\boldsymbol{\varepsilon}_j$.*

Let $f_{\bar{l}}$ be diagnosable for $1 \le \bar{l} \le r$ and estimated by means of the dynamics of $\hat{\mathbf{f}}_{\bar{l}}$. Let $\hat{u}_l$ be the solution to

$$-L_j(\hat{\mathbf{e}}, \mathbf{y}_R, \hat{\mathbf{u}}, \hat{u}_l^{(\gamma_l)}, \hat{\mathbf{f}}) - y_{Rj}^{(n)} = -\sum_{i=1}^n a_{ij}\hat{e}_{ij}$$

Then, closed-loop system (4.30) with control $\hat{u} = (\hat{u}_1,\dots,\hat{u}_p)$ is asymptotically stable.

Proof Consider the following Lyapunov function:

$$V(\hat{\mathbf{e}}_j, \boldsymbol{\varepsilon}_j, \tilde{f}_{\bar{l}}) = V_1(\hat{\mathbf{e}}_j) + V_2(\boldsymbol{\varepsilon}_j) + V_3(\tilde{f}_{\bar{l}}) \tag{4.31}$$

with

$$V_1(\hat{\mathbf{e}}_j) = \hat{\mathbf{e}}_j^T P \hat{\mathbf{e}}_j \qquad V_2(\boldsymbol{\varepsilon}_j) = \boldsymbol{\varepsilon}_j^T S_\infty \boldsymbol{\varepsilon}_j \qquad V_3(\tilde{f}_{\bar{l}}) = \tilde{f}_{\bar{l}}^T I \tilde{f}_{\bar{l}}$$

where $\tilde{f}_{\bar{l}} = f_{\bar{l}} - \hat{f}_{\bar{l}}$. Besides, P is the solution to $F^T P + P F = -I$. Then P is positive-definite and define $\|x\|_P = \sqrt{x^T P x}$. Let S_∞ be the solution to (4.23), then S_∞ is positive-definite and denote $\|x\|_{S_\infty} = \sqrt{x^T S_\infty x}$.

Now, taking the derivative with respect to time of the first term of (4.31):

$$\dot{V}_1(\hat{\mathbf{e}}_j) = \hat{\mathbf{e}}_j^T P \dot{\hat{\mathbf{e}}}_j + \dot{\hat{\mathbf{e}}}_j^T P \hat{\mathbf{e}}_j \le -\alpha \hat{\mathbf{e}}_j^T P \hat{\mathbf{e}}_j - 2\hat{\mathbf{e}}_j^T P S_\infty^{-1} C^T C \boldsymbol{\varepsilon}_j \tag{4.32}$$

where $\alpha = 1/\lambda_{\max}(P)$. Noting that

$$\left\|\hat{\mathbf{e}}_j^T P S_\infty^{-1} C^T C S_\infty^{-1} S_\infty \boldsymbol{\varepsilon}_j\right\| \le \rho\left(\theta\right)\left\|\hat{\mathbf{e}}_j\right\|_P \left\|\boldsymbol{\varepsilon}_j\right\|_{S_\infty}$$

with $\rho(\theta) = \left\|S_\infty^{-1} C^T C S_\infty^{-1}\right\|$, then the next inequality is obtained:

$$\dot{V}_1(\hat{\mathbf{e}}_j) \le -\left(\alpha\left\|\hat{\mathbf{e}}_j\right\|_P + 2\rho\left(\theta\right)\left\|\boldsymbol{\varepsilon}_j\right\|_{S_\infty}\right)\left\|\hat{\mathbf{e}}_j\right\|_P \tag{4.33}$$

Let d_1, d_2 be positive numbers such that $\left\| \hat{\mathbf{e}}_j \right\|_P \geq d_1 \left\| \hat{\mathbf{e}}_j \right\|$ and $\left\| \boldsymbol{\varepsilon}_j \right\|_{S_\infty} \geq d_2 \left\| \boldsymbol{\varepsilon}_j \right\|$. Thus the inequality is rewritten as:

$$\dot{V}_1(\hat{\mathbf{e}}_j) \leq - \left(\alpha d_1 \left\| \hat{\mathbf{e}}_j \right\| + 2\rho\,(\theta)\, d_2 \left\| \boldsymbol{\varepsilon}_j \right\| \right) \left\| \hat{\mathbf{e}}_j \right\|_P \tag{4.34}$$

and this yields to

$$\dot{V}_1(\hat{\mathbf{e}}_j) \leq 0$$

Taking the derivative with respect to time of the second term of (4.31):

$$\dot{V}_2(\boldsymbol{\varepsilon}_j) = \boldsymbol{\varepsilon}_j^T S_\infty \dot{\boldsymbol{\varepsilon}}_j + \dot{\boldsymbol{\varepsilon}}_j^T S_\infty \boldsymbol{\varepsilon}_j \leq -\theta \left\| \boldsymbol{\varepsilon}_j \right\|_{S_\infty}^2 + 2 \left\| \boldsymbol{\varepsilon}_j \right\|_{S_\infty} \left\| \Phi_j(\boldsymbol{\varepsilon}, \hat{\mathbf{e}}) \right\|_{S_\infty} \tag{4.35}$$

where it was taken into account (4.23), the Cholesky decomposition and the fact that $\boldsymbol{\varepsilon}_j^T C^T C \boldsymbol{\varepsilon}_j \geq 0$. Besides, noting that $\Phi_j(\boldsymbol{\varepsilon}, \hat{\mathbf{e}})$ is differentiable, by the Lipschitz property:

$$\left\| \Phi_j(\boldsymbol{\varepsilon}, \hat{\mathbf{e}}) \right\|_{S_\infty} \leq \lambda \left\| \boldsymbol{\varepsilon}_j \right\|_{S_\infty} \qquad 1 \leq j \leq p$$

thus:

$$\dot{V}_2(\boldsymbol{\varepsilon}_j) \leq - (\theta - 2\lambda) \left\| \boldsymbol{\varepsilon}_j \right\|_{S_\infty}^2 \tag{4.36}$$

with $\lambda < \theta/2$.

From (4.36) the following inequality is obtained:

$$\frac{d \left(\left\| \boldsymbol{\varepsilon}_j \right\|_{S_\infty}^2 \right)}{dt} \leq - (\theta - 2\lambda) \left\| \boldsymbol{\varepsilon}_j \right\|_{S_\infty}^2 \tag{4.37}$$

that yields:

$$\left\| \boldsymbol{\varepsilon}_j \right\|_{S_\infty} \leq -e^{-\gamma t} \left\| \boldsymbol{\varepsilon}_j(0) \right\|_{S_\infty} \tag{4.38}$$

with $\gamma = \theta/2 - \lambda$.

Similarly, from (4.33) the next equation is achieved:

$$\frac{d \left(\left\| \hat{\mathbf{e}}_j \right\|_P^2 \right)}{dt} \leq - \left(\alpha \left\| \hat{\mathbf{e}}_j \right\|_P + 2\rho\,(\theta) \left\| \boldsymbol{\varepsilon}_j \right\|_{S_\infty} \right) \left\| \hat{\mathbf{e}}_j \right\|_P \tag{4.39}$$

this yields:

$$\left\| \hat{\mathbf{e}}_j \right\|_P \leq A e^{-\frac{\alpha}{2}t} + B e^{-\gamma t} \tag{4.40}$$

with

$$A = \left\| \hat{e}_j(0) \right\|_P - B$$

$$B = - \frac{\rho\,(\theta) \left\| \varepsilon_j(0) \right\|_{S_\infty}}{\alpha/2 - \gamma}$$

Finally, taking the derivative with respect to time of the third term of (4.31):

$$V_3(\tilde{f}_{\bar{l}}) = \tilde{f}_{\bar{l}}^T I \tilde{f}_{\bar{l}} = \tilde{f}_{\bar{l}}^2 \tag{4.41}$$

$$\dot{V}_3(\tilde{f}_{\bar{l}}) = 2\tilde{f}_{\bar{l}}\dot{\tilde{f}}_{\bar{l}} \le 2\tilde{f}_{\bar{l}}^T \left(\frac{N_{\bar{l}}}{k_{\bar{l}}} - \tilde{f}_{\bar{l}} \right) \tag{4.42}$$

and the condition $N_{\bar{l}}/k_{\bar{l}} \to 0$ with $t \to \infty$ is imposed, then

$$\dot{V}_3(\tilde{f}_{\bar{l}}) \le -2\tilde{f}_{\bar{l}}^2 \tag{4.43}$$

Thus, the system (4.1) with control $\hat{u} = (\hat{u}_1, \ldots, \hat{u}_p)$ is asymptotically stable, for $1 \le j \le p, 1 \le l \le m$ and $1 \le \bar{l} \le r$. $\qquad\square$

The following remark shows the relation of the convergence velocity between the tracking error and the observation error.

Remark 4.5 It can be seen from (4.40) that

$$\left\| \hat{\mathbf{e}}_j \right\|_P \le (A + B) e^{-\min\left\{ \frac{\alpha}{2}, \gamma \right\} t} \tag{4.44}$$

Selecting the condition $\theta/2 - \lambda = \gamma > \alpha/2$, θ can be chosen such that for a fixed value:

$$\left\| \hat{\mathbf{e}}_j \right\|_P \le (A + B) e^{-\frac{\alpha}{2} t} \tag{4.45}$$

This implies that the observation error $\boldsymbol{\varepsilon}_j$ converges faster than the estimated tracking error $\hat{\mathbf{e}}_j$.

Note that the following remark establishes the stability region obtained considering a measurement noise in the output, using the uniform ultimate boundedness theorem [5].

Remark 4.6 Consider the above problem but with a bounded deterministic noise δ such that $\|\delta\| \le \delta^+, \delta^+ > 0$, with corrupted measurement, i.e. $y + \delta$. In this case the observation error is defined as $\varepsilon_j = \hat{\mathbf{e}}_j - \mathbf{e}_j - \boldsymbol{\delta}_j$, where $\boldsymbol{\delta}_j = col\left(\delta_j \ 0 \ldots 0 \right)$ and $\delta_j \in \mathbb{R}$. In this case, the dynamics of the HGO (4.26) is given by:

$$\dot{\hat{\mathbf{e}}}_j = E\hat{\mathbf{e}}_j + \hat{\varphi}_j \left(\hat{\mathbf{e}}, \mathbf{y}_R, y_{Rj}^{(n)}, \hat{\mathbf{u}}, \hat{\mathbf{f}} \right) - S_\infty^{-1} C^T C(\hat{\mathbf{e}}_j - \mathbf{e}_j - \boldsymbol{\delta}_j) \tag{4.46}$$

$$= E\hat{\mathbf{e}}_j + \hat{\varphi}_j \left(\hat{\mathbf{e}}, \mathbf{y}_R, y_{Rj}^{(n)}, \hat{\mathbf{u}}, \hat{\mathbf{f}} \right) - S_\infty^{-1} C^T C(\hat{\mathbf{e}}_j - \mathbf{e}_j) + S_\infty^{-1} C^T C\boldsymbol{\delta}_j$$

and thus the observation dynamics is written as follows:

$$\dot{\varepsilon}_j = E\varepsilon - S_\infty^{-1} C^T C\varepsilon + \hat{\varphi}_j \left(\hat{\mathbf{e}}, \mathbf{y}_R, y_{Rj}^{(n)}, \hat{\mathbf{u}}, \hat{\mathbf{f}} \right) - \varphi_j \left(\mathbf{e}, \mathbf{y}_R, y_{Rj}^{(n)}, \hat{\mathbf{u}}, \hat{\mathbf{f}} \right)$$

$$+ S_\infty^{-1} C^T C\boldsymbol{\delta}_j = E_\theta \varepsilon_j + \Phi_j \left(\varepsilon, \hat{\mathbf{e}} \right) + W_j \tag{4.47}$$

where $W_j = S_\infty^{-1} C^T C \boldsymbol{\delta}_j$.

Hence, the derivative of the second term of the Lyapunov function (4.31), i.e. $V_2(\boldsymbol{\varepsilon}_j) = \boldsymbol{\varepsilon}_j^T S_\infty \boldsymbol{\varepsilon}_j$, is:

$$
\begin{aligned}
\dot{V}_2(\varepsilon_j) &= \dot{\varepsilon}_j^T S_\infty \varepsilon_j + \varepsilon_j^T S_\infty \dot{\varepsilon}_j \\
&\leq -\theta \varepsilon_j^T S_\infty \varepsilon_j + 2\varepsilon_j^T S_\infty \Phi_j + 2\varepsilon_j^T S_\infty W_j \\
&\leq -\theta \left\| \varepsilon_j \right\|_{S_\infty}^2 + 2 \left\| \varepsilon_j \right\|_{S_\infty} \left\| \Phi_j \right\|_{S_\infty} + 2 \left\| \varepsilon_j \right\|_{S_\infty} \left\| W_j \right\|_{S_\infty} \\
&\leq -(\theta - 2\lambda) \left\| \varepsilon_j \right\|_{S_\infty}^2 + 2 \left\| \varepsilon_j \right\|_{S_\infty} \left\| W_j \right\|_{S_\infty}
\end{aligned}
\tag{4.48}
$$

Since the term W_j is given by [16]

$$
W_j = S_\infty^{-1} C^T C \boldsymbol{\delta}_j =
\begin{pmatrix}
-n\theta & 0 \dots 0 \\
-\frac{n(n-1)}{2!}\theta^2 & 0 \dots 0 \\
-\frac{n(n-1)(n-2)}{2!}\theta^3 & 0 \dots 0 \\
\vdots & \vdots \ddots \vdots \\
-\frac{n(n-1)}{2!}\theta^{n-2} & 0 \dots 0 \\
-n\theta^{n-1} & 0 \dots 0 \\
-\theta^n & 0 \dots 0
\end{pmatrix}
\begin{pmatrix}
\delta_j \\
0 \\
\vdots \\
0
\end{pmatrix}
$$

$$
W_j =
\begin{pmatrix}
-n\theta \delta_j \\
-\frac{n(n-1)}{2!}\theta^2 \delta_j \\
-\frac{n(n-1)(n-2)}{2!}\theta^3 \delta_j \\
\vdots \\
-\frac{n(n-1)}{2!}\theta^{n-2} \delta_j \\
-n\theta^{n-1} \delta_j \\
-\theta^n \delta_j
\end{pmatrix}
\tag{4.49}
$$

given that the noise is deterministic and bounded, i.e. $\left\| \delta_j \right\| \leq \delta^+$, $n \in \mathbb{Z}^+$ is finite and $\theta > 0$, then $\exists\, \Gamma > 0$ finite, such that $\left\| W_j \right\|_{S_\infty} \leq \Gamma$. Therefore:

$$
\dot{V}_2(\varepsilon_j) \leq -(\theta - 2\lambda) \left\| \varepsilon_j \right\|_{S_\infty}^2 + 2\Gamma \left\| \varepsilon_j \right\|_{S_\infty}
\tag{4.50}
$$

Now, applying the Rayleigh–Ritz inequality:

$$
\dot{V}_2(\varepsilon_j) \leq -(\theta - 2\gamma) \lambda_{\min}(S_\infty) \left\| \varepsilon_j \right\|^2 + 2\Gamma \sqrt{\lambda_{\max}(S_\infty)} \left\| \varepsilon_j \right\|
\tag{4.51}
$$

Finally, applying the uniform ultimate boundedness theorem [5] it is concluded that ε_j is bounded uniformly by any initial state $\varepsilon_j(0)$ and remains in a compact set $B_b = \{\varepsilon_j : \|\varepsilon_j\| \le b, \ b > 0\}$, where the ultimate bound is defined as

$$b = \sqrt{\frac{\lambda_{\max}(S_\infty)}{\lambda_{\min}(S_\infty)}} \left(\frac{2\Gamma \sqrt{\lambda_{\max}(S_\infty)}}{(\theta - 2\gamma)\lambda_{\min}(S_\infty)} \right) \tag{4.52}$$

Furthermore, from a similar analysis, the derivative of the first term of the Lyapunov function (4.31), i.e. $V_1(\hat{\mathbf{e}}_j) = \hat{\mathbf{e}}_j^T P \hat{\mathbf{e}}_j$ is:

$$\dot{V}_1(\hat{\mathbf{e}}_j) = \hat{\mathbf{e}}_j^T P \dot{\hat{\mathbf{e}}}_j + \dot{\hat{\mathbf{e}}}_j^T P \hat{\mathbf{e}}_j$$
$$\le - \left(\alpha d_1 \|\hat{\mathbf{e}}_j\| + 2\rho(\theta) d_2 \|\varepsilon_j\| - 2\Gamma \right) \|\hat{\mathbf{e}}_j\|_P \tag{4.53}$$

and we obtain the following ultimate bound for $\hat{\mathbf{e}}_j$:

$$b = \sqrt{\frac{(\lambda_{\max}(P))^3}{\lambda_{\min}(P) d_1^2}} \left(-\frac{4\rho(\theta) d_2 \Gamma \sqrt{\lambda_{\max}(S_\infty)}}{(\theta - 2\gamma)\lambda_{\min}(S_\infty)} + 2\Gamma \right) \tag{4.54}$$

So, in the presence of measurement noise, ε_j and $\hat{\mathbf{e}}_j$ are uniform ultimate bounded; the effect of this can be seen in the experimental results (see Fig. 4.9).

4.4   Application

4.4.1   Numerical Example

Consider the following nonlinear system

$$\dot{x}_1 = x_1 x_2 + f + u$$
$$\dot{x}_2 = x_1 \tag{4.55}$$
$$\dot{x}_3 = x_3 f + u$$
$$y = x_2$$

Choosing $\eta_1 = y$, the system can be transformed into the MMGOCF (4.11):

$$\dot{\eta}_1 = \eta_2$$
$$\dot{\eta}_2 = \eta_3 \tag{4.56}$$
$$\dot{\eta}_3 = (\eta_2)^2 + \eta_2 (\eta_1)^2 + \eta_1 u + \eta_1 f + \dot{u} + \dot{f}$$

Now, the tracking error HGO (4.22) is built, where

$$E = \begin{pmatrix} 0 & 1 & 0 \\ 0 & 0 & 1 \\ 0 & 0 & 0 \end{pmatrix} \qquad C = \begin{pmatrix} 1 & 0 & 0 \end{pmatrix}$$

and matrix S_∞^{-1} is chosen as:

$$S_\infty^{-1} = \begin{pmatrix} 3\theta & 3\theta^2 & \theta^3 \\ 3\theta^2 & 5\theta^3 & 2\theta^4 \\ \theta^3 & 2\theta^4 & \theta^5 \end{pmatrix}$$

Defining $e_1 = \eta_1 - y_R$ and $e_1^{(i-1)} = e_i$, the tracking error HGO is written as:

$$\dot{\hat{e}}_1 = \hat{e}_2 - 3\theta(\hat{e}_1 - e_1) \tag{4.57}$$
$$\dot{\hat{e}}_2 = \hat{e}_3 - 3\theta^2(\hat{e}_1 - e_1)$$
$$\dot{\hat{e}}_3 = (\hat{e}_2 + \dot{y}_R)^2 + (\hat{e}_2 + \dot{y}_R)(\hat{e}_1 + y_R)^2 + (\hat{e}_1 + y_R)\hat{u} + (\hat{e}_1 + y_R)\hat{f} + \dot{\hat{u}} + \dot{\hat{f}}$$
$$-\dddot{y}_R - \theta^3(\hat{e}_1 - e_1) = -\sum_{i=1}^{3} a_i \hat{e}_i$$

From system (4.57), the dynamical equation of the controller (4.25) is obtained as:

$$\dot{\hat{u}} = -\sum a_i \hat{e}_i - (\hat{e}_2 + \dot{y}_R)^2 - (\hat{e}_2 + \dot{y}_R)(\hat{e}_1 + y_R)^2 - (\hat{e}_1 + y_R)\hat{u}$$
$$- (\hat{e}_1 + y_R)\hat{f} - \dot{\hat{f}} + \dddot{y}_R \tag{4.58}$$

Defining $\dot{\hat{u}} = \hat{u}_1$, the dynamics of the controller is:

$$\hat{u}_1 = -(\hat{e}_2 + \dot{y}_R)^2 - (\hat{e}_2 + \dot{y}_R)(\hat{e}_1 + y_R)^2 - (\hat{e}_1 + y_R)\hat{u}$$
$$- (\hat{e}_1 + y_R)\hat{f} - \dot{\hat{f}} + \dddot{y}_R \tag{4.59}$$

Furthermore, it can be seen that system (4.55) is diagnosable, because the following polynomial can be obtained:

$$f - \ddot{y} + \dot{y}y + u = 0 \tag{4.60}$$

thus an observer for estimating f can be built based on Eq. (4.6). For this, the following ROO is proposed:

$$\dot{\gamma}_2 = -k_2(\hat{x}_1 y + \hat{u} + \gamma_2 + k_2 \hat{x}_1) \tag{4.61}$$
$$\hat{f} = \gamma_2 + k_2 \hat{x}_1$$

where $\hat{x}_1$ is the estimation of $\hat{y}$, which is obtained with

$$\dot{\gamma}_1 = -k_1 (\gamma_1 + k_1 y) \tag{4.62}$$
$$\hat{x}_1 = \gamma_1 + k_1 y$$

Defining $\hat{f} = \hat{f}_1$, the dynamics of the fault is:

$$\dot{\hat{f}}_1 = -k_2(\hat{x}_1 y + \hat{u} + \gamma_2 + k_2 \hat{x}_1) + k_2(-k_1 (\gamma_1 + k_1 y) + k_1 \hat{x}_1) \tag{4.63}$$

Simulations were made for this system over $15\,\mathrm{s}$, using the time-variant reference $y_R = 0.1 \sin(t)$ and the fault $f = 50 \left[1 + \sin(0.2 t e^{-0.5t})\right] \mathscr{U}(t-5)$, where $\mathscr{U}(t)$ is the step function. The design parameters were chosen as $\theta = 20$, $a_1 = 8000$, $a_2 = 1200$, $a_3 = 60$, $k_1 = k_2 = 1$.

Figure 4.1 shows how the output y follows the reference y_R; it can be seen that approximately after one second the output follows the sinusoidal reference. However, the effect of f, which begins at $5\,\mathrm{s}$, can be appreciated before it is eliminated.

Figure 4.2 shows the fault-tolerant control signal used. It can be appreciated that the controller uses less energy when the fault appears; this is due to the nature of the fault, whose amplitude has a positive value. The controller seeks to compensate it, and the elimination of the effects of the fault can be seen in the output tracking graph.

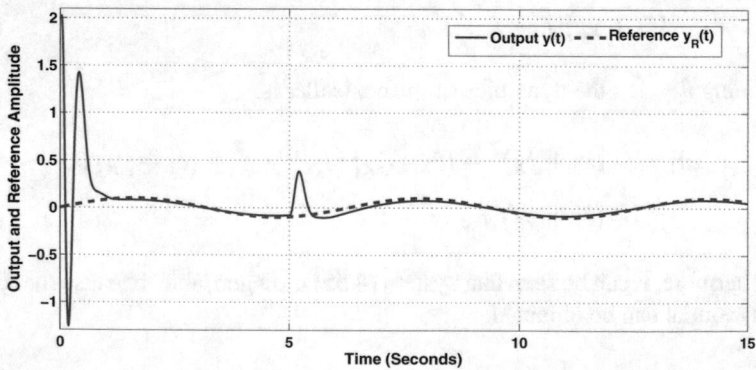

Fig. 4.1 Output tracking of the numerical example

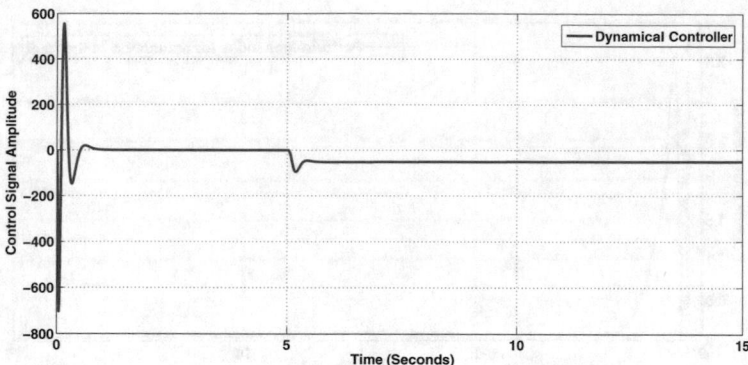

Fig. 4.2 Dynamical controller of the numerical example

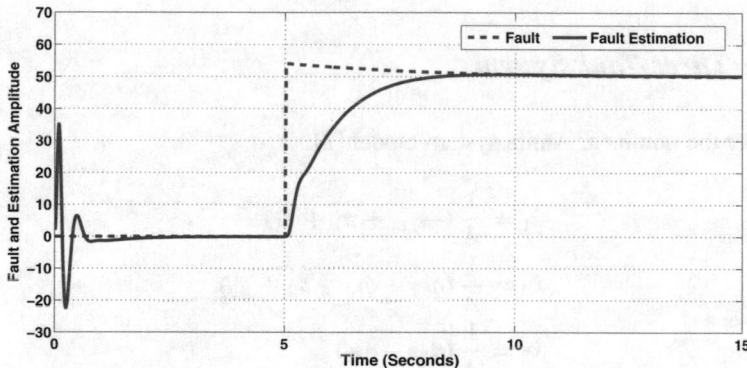

Fig. 4.3 Fault estimation of the numerical example

The fault f and its estimation $\hat{f}$ are shown in Fig. 4.3. It can be seen that the ROO transient part ends quickly, and 5 s after the fault appears, the estimation follows the real signal.

Finally, Fig. 4.4 shows the performance index of the ROO proposed to estimate the fault, which was evaluated using the following cost function:

$$J_t = \frac{1}{t + \varepsilon} \int_0^t \left\| \tilde{f}_l \right\|^2 dt \qquad (4.64)$$

where $\varepsilon = 0.0001$. This parameter is used to avoid singularities when $t = 0$, and is chosen sufficiently small so that it does not alter significantly the value of the index.

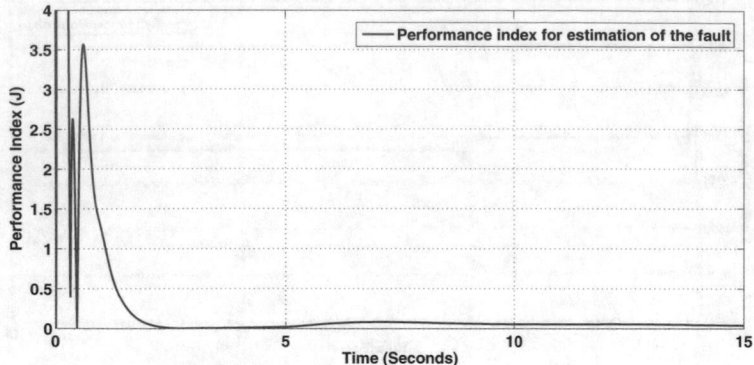

Fig. 4.4 Performance evaluation for the fault estimation of the numerical example

4.4.2   Three-Tank System

Consider the nonlinear Amira system model [2]:

$$\dot{x}_1 = \frac{1}{A}\left(-q_{13} + u_1 + f_1\right) \tag{4.65}$$

$$\dot{x}_2 = \frac{1}{A}\left(q_{32} - q_{20} + u_2 + f_2\right)$$

$$\dot{x}_3 = \frac{1}{A}\left(q_{13} - q_{32}\right)$$

$$y_1 = x_2$$

$$y_2 = x_3$$

with

$$q_{13} = a_1 S\sqrt{2g(x_1 - x_3)}$$

$$q_{32} = a_3 S\sqrt{2g(x_3 - x_2)}$$

$$q_{20} = a_2 S\sqrt{2gx_2}$$

In this system, $u_i = q_i$, $i = 1, 2$ are the manipulable input flows, $x_i = h_i$, $i = 1, 2, 3$ are the levels of each tank, A is the cross section of the tanks, and the terms q_{ij} represent the water flow from tank i to tank j. S is the cross-sectional area of the pipe that interconnects each tank and the unknown parameters a_i, $i = 1, 2, 3$ are the output flow coefficients.

It is important to mention that the system has four operation regions, and the region considered here is $h_1 > h_3 > h_2 > 0$. To strengthen the information, the characteristics and variables of the system and how the system operates on the desired region

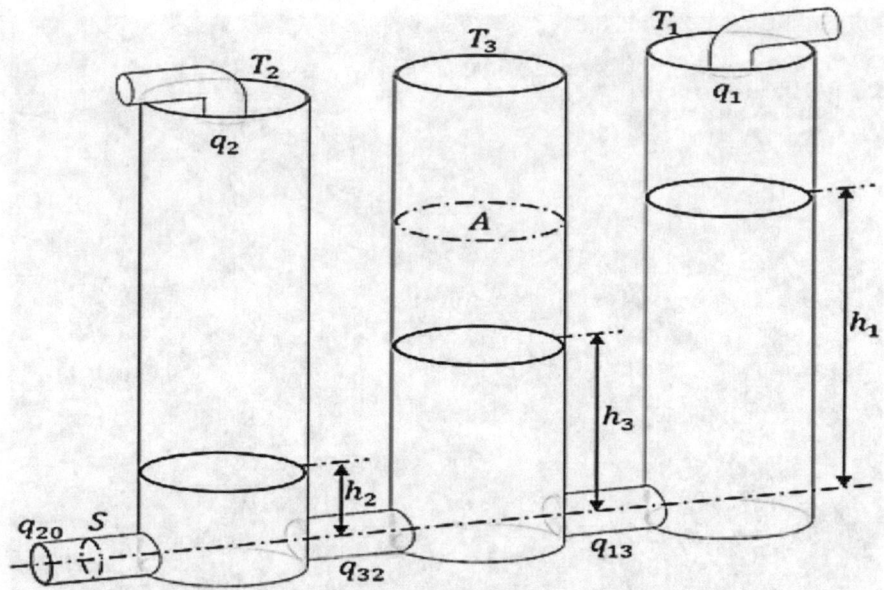

Fig. 4.5 Schematic diagram of the Amira DTS200 working in the region $h_1 > h_3 > h_2 > 0$

are shown on Fig. 4.5. On Fig. 4.6, a picture of the real system used is shown, which is found in the Automatic Control Department, Cinvestav-IPN, México.

From the model (4.65), it can be seen that x_1 is not available for measurement, however it is algebraically observable according to Definition 4.2, which allows to write the following algebraic polynomial:

$$x_1 - y_2 - \frac{1}{2ga_1^2S^2}\left(A\dot{y}_2 + a_3S\sqrt{2g(y_2 - y_1)}\right)^2 = 0. \tag{4.66}$$

4.4.2.1 Parameter Estimation

As it is mentioned above, the flow transfer coefficients a_1, a_2 and a_3 are not known, however it can be verified that they are algebraically observable, that is, they satisfy differential algebraic equations with coefficients in $k\langle u, y \rangle$ (see definition of algebraically observable parameter [16]). Thus, from (4.65) the following expressions, which are defined in the region of interest ($h_1 > h_3 > h_2 > 0$) are obtained:

Fig. 4.6 Amira DTS200 three-tank system benchmark

$$a_1 = \frac{q_1 - A\dot{x}_1}{S\sqrt{2g(x_l - x_3)}} \tag{4.67}$$

$$a_2 = \frac{q_1 + q_2 - A(\dot{x}_1 + \dot{x}_2 + \dot{x}_3)}{S\sqrt{2g(x_2)}} \tag{4.68}$$

$$a_3 = \frac{q_1 - A(\dot{x}_1 + \dot{x}_3)}{S\sqrt{2g(x_3 - x_2)}} \tag{4.69}$$

Thus, if the measurements of the inputs q_1 and q_2 and the state variables x_1, x_2 and x_3 from the nominal model (without the presence of faults) are available, it is possible to estimate their time derivatives and from (4.67)–(4.69) the estimates of the uncertain parameters a_1, a_2 and a_3 can be obtained.

4.4.2.2 Experimental Results

Now, the proposed method is applied. Firstly, the change of variables $\eta_1^j = y_j$ is selected, so the system (4.65) can be transformed into the MMGOCF (4.11). For $y_1 = \eta_{11}$, the subsystem is:

$$\dot{\eta}_{11} = \eta_{21} \tag{4.70}$$

$$\dot{\eta}_{21} = \eta_{31}$$

$$\dot{\eta}_{31} = \frac{1}{A}\left(a_3 Sg^2 \frac{2(\eta_{12} - \eta_{11})(\eta_{32} - \eta_{31}) - (\eta_{22} - \eta_{21})^2}{\sqrt{(2g(\eta_{12} - \eta_{11}))^3}} \right.$$

$$\left. - a_2 Sg^2 \frac{2\eta_{11}\eta_{31} - \eta_{21}^2}{\sqrt{(2g\eta_{11})^3}} + \ddot{u}_2 + \ddot{f}_2 \right)$$

and for $y_2 = \eta_{12}$:

$$\dot{\eta}_{12} = \eta_{22} \tag{4.71}$$

$$\dot{\eta}_{22} = \eta_{32}$$

$$\dot{\eta}_{32} = \frac{1}{A^2}\left(a_1^2 S^2 g^3 \frac{2(\dot{x}_1 - \eta_{22})(x_1 - \eta_{12})}{(2g(x_1 - \eta_{12}))^2} + a_1 Sg^2 \frac{2(x_1 - \eta_{12})}{\sqrt{(2g(x_1 - \eta_{12}))^3}} \left(\dot{u}_1 + \dot{f}_1 \right) \right)$$

$$- \frac{1}{A}\left(a_1 Sg^2 \frac{2\,(\eta_{32})\,(x_1 - \eta_{12}) + (\dot{x}_1 - \eta_{22})^2}{\sqrt{(2g(x_1 - \eta_{12}))^3}} \right. \tag{4.72}$$

$$\left. + a_3 Sg^2 \frac{2(\eta_{12} - \eta_{11})(\eta_{32} - \eta_{31}) - (\eta_{22} - \eta_{21})^2}{\sqrt{(2g(\eta_{12} - \eta_{11}))^3}} \right)$$

These subsystems allow the construction of the tracking error HGO (4.26), where the matrices E and C are defined as

$$E = \begin{pmatrix} 0 & 1 & 0 \\ 0 & 0 & 1 \\ 0 & 0 & 0 \end{pmatrix} \qquad C = \begin{pmatrix} 1 & 0 & 0 \end{pmatrix}$$

and S_∞^{-1} is chosen as follows:

$$S_\infty^{-1} = \begin{pmatrix} 3\theta & 3\theta^2 & \theta^3 \\ 3\theta^2 & 5\theta^3 & 2\theta^4 \\ \theta^3 & 2\theta^4 & \theta^5 \end{pmatrix}$$

Defining $e_{11} = \eta_{11} - y_{R1}$, $e_{12} = \eta_{12} - y_{R2}$, $e_1^{(i-1)} = e_{i1}$ and $e_2^{(i-1)} = e_{i2}$, the HGO for $y_1 = \eta_{11}$ is:

$$\dot{\hat{e}}_{11} = \hat{e}_{21} - 3\theta_1(\hat{e}_{11} - e_{11}) \tag{4.73}$$

$$\dot{\hat{e}}_{21} = \hat{e}_{31} - 3\theta_1^2(\hat{e}_{11} - e_{11})$$

$$\dot{\hat{e}}_{31} = \frac{a_3 S \sqrt{g}}{A} \left(\frac{1}{\sqrt{(2((\hat{e}_{12} + y_{R2}) - (\hat{e}_{11} + y_{R1})))^3}} \right) \times$$

$$\left(2\left((\hat{e}_{12} + y_{R2}) - (\hat{e}_{11} + y_{R1})\right) (\hat{e}_{32} - \hat{e}_{31}) - (\hat{e}_{22} - \hat{e}_{21})^2 \right)$$

$$- \frac{a_2 S \sqrt{g}}{A} \left(\frac{2 (\hat{e}_{11} + y_{R1}) \hat{e}_{31} - \hat{e}_{21}^2}{\sqrt{(2 (\hat{e}_{11} + y_{R1}))^3}} \right) + \frac{1}{A} \left(\ddot{\hat{u}}_2 + \ddot{\hat{f}}_2 \right) - \theta_1^3 (\hat{e}_{11} - e_{11}) \;\; = \;\; - \sum_{i=1}^{3} s_i \hat{e}_{i1}$$

and for $y_2 = \eta_{12}$:

$$\dot{\hat{e}}_{12} = \hat{e}_{22} - 3\theta_2 (\hat{e}_{12} - e_{12}) \tag{4.74}$$

$$\dot{\hat{e}}_{22} = \hat{e}_{32} - 3\theta_2^2 (\hat{e}_{12} - e_{12})$$

$$\dot{\hat{e}}_{32} = - \frac{a_3 S \sqrt{g}}{A} \left(\frac{1}{\sqrt{(2((\hat{e}_{12} + y_{R2}) - (\hat{e}_{11} + y_{R1})))^3}} \right) \times$$

$$\left(2\left((\hat{e}_{12} + y_{R2}) - (\hat{e}_{11} + y_{R1})\right) (\hat{e}_{32} - \hat{e}_{31}) - (\hat{e}_{22} - \hat{e}_{21})^2 \right)$$

$$- \frac{a_1 S \sqrt{g}}{A} \left(\frac{2\hat{e}_{32}(\hat{x}_1 - (\hat{e}_{12} + y_{R2})) + \left(\dot{\hat{x}}_1 - \hat{e}_{22} \right)^2}{\sqrt{(2(\hat{x}_1 - (\hat{e}_{12} + y_{R2})))^3}} \right)$$

$$+ \frac{a_1 S \sqrt{g}}{A^2} \left(\frac{1}{\sqrt{2(\hat{x}_1 - (\hat{e}_{12} + y_{R2}))}} \right) \left(\dot{\hat{u}}_1 + \dot{\hat{f}}_1 \right)$$

$$+ \frac{a_1^2 S^2 g}{2A^2} \left(\frac{\dot{\hat{x}}_1 - \hat{e}_{22}}{\hat{x}_1 - (\hat{e}_{12} + y_{R2})} \right) - \theta_2^3 (\hat{e}_{12} - e_{12}) \;\; = \;\; - \sum_{i=1}^{3} t_i \hat{e}_{i2}$$

Given that the references for this system are constant, their derivatives have been neglected in these equations. Consequently, from subsystem (4.74) the dynamics of $\hat{u}_1$ is:

$$\dot{\hat{u}}_1 = \frac{A^2 \sqrt{2(\hat{x}_1 - (\hat{e}_{12} + y_{R2}))}}{a_1 S \sqrt{g}} \left(- \sum_{i=1}^{3} t_i \hat{e}_{i2} \right) - \frac{a_1 S \sqrt{g} \left(\dot{\hat{x}}_1 - \hat{e}_{22} \right)}{\sqrt{2(\hat{x}_1 - (\hat{e}_{12} + y_{R2}))}}$$

$$+ \frac{2A\hat{e}_{32}(\hat{x}_1 - (\hat{e}_{12} + y_{R2})) + A \left(\dot{\hat{x}}_1 - \hat{e}_{22} \right)^2}{2(\hat{x}_1 - (\hat{e}_{12} + y_{R2}))} \tag{4.75}$$

$$+ \frac{a_3 A \sqrt{(\hat{x}_1 - (\hat{e}_{12} + y_{R2}))}}{2a_1} \left(\frac{1}{\sqrt{((\hat{e}_{12} + y_{R2}) - (\hat{e}_{11} + y_{R1}))^3}} \right) \times$$

$$\left(2\left((\hat{e}_{12}+y_{R2})-(\hat{e}_{11}+y_{R1})\right)(\hat{e}_{32}-\hat{e}_{31})-(\hat{e}_{22}-\hat{e}_{21})^2\right)-\dot{\hat{f}}_1$$

and from subsystem (4.73), the dynamics of $\hat{u}_2$ is given by:

$$
\begin{aligned}
\ddot{\hat{u}}_2 = {} & A\left(-\sum_{i=1}^{3}s_i\hat{e}_{i1}\right)+a_2 S\sqrt{g}\left(\frac{2(\hat{e}_{11}+y_{R1})\hat{e}_{31}-(\hat{e}_{21})^2}{\sqrt{(2(\hat{e}_{11}+y_{R1}))^3}}\right) \\
& -a_3 S\sqrt{g}\left(\frac{1}{\sqrt{(2((\hat{e}_{12}+y_{R2})-(\hat{e}_{11}+y_{R1})))^3}}\right)\times \\
& \left(2\left((\hat{e}_{12}+y_{R2})-(\hat{e}_{11}+y_{R1})\right)(\hat{e}_{32}-\hat{e}_{31})-(\hat{e}_{22}-\hat{e}_{21})^2\right)-\ddot{\hat{f}}_2
\end{aligned}
\tag{4.76}
$$

Defining $\hat{u}_1 = \hat{u}_{11}$, $\hat{u}_2 = \hat{u}_{12}$ and $\dot{\hat{u}}_2 = \hat{u}_{22}$, the chains of integrators of the controllers are:

$$
\begin{aligned}
\dot{\hat{u}}_{11} = {} & \frac{A^2\sqrt{2T_3}}{a_1 S\sqrt{g}}\left(\dddot{y}_{R2}-\sum_{i=1}^{3}t_i\hat{e}_{i2}\right)-\frac{a_1 S\sqrt{g}\dot{T}_3}{\sqrt{2T_3}} \\
& +\frac{a_3 A\sqrt{T_3}}{2a_1}\left(\frac{2T_1\ddot{T}_1-(\dot{T}_1)^2}{\sqrt{(T_1)^3}}\right) \\
& +\frac{2A(\hat{e}_{32}+\ddot{y}_{R2})T_3+A(\dot{T}_3)^2}{2T_3}-\dot{\hat{f}}_1
\end{aligned}
\tag{4.77}
$$

$$\dot{\hat{u}}_{12} = \hat{u}_{22}$$

$$
\begin{aligned}
\dot{\hat{u}}_{22} = {} & A\left(\dddot{y}_{R1}-\sum_{i=1}^{3}s_i\hat{e}_{i1}\right)-a_3 S\sqrt{g}\left(\frac{2T_1\ddot{T}_1-(\dot{T}_1)^2}{\sqrt{(2T_1)^3}}\right) \\
& +a_2 S\sqrt{g}\left(\frac{2T_2\ddot{T}_2-(\dot{T}_2)^2}{\sqrt{(2T_2)^3}}\right)-\ddot{\hat{f}}_2
\end{aligned}
$$

where

$$
\begin{aligned}
T_1 &= (\hat{e}_{12}+y_{R2})-(\hat{e}_{11}+y_{R1}) \\
T_2 &= (\hat{e}_{11}+y_{R1}) \\
T_3 &= \hat{x}_1-(\hat{e}_{12}+y_{R2})
\end{aligned}
$$

Furthermore, it can be seen from (4.65) that f_1 and f_2 are diagnosable, because the following polynomials can be obtained:

$$f_1 - A\dot{x}_1 - a_1 S\sqrt{2g(x_1 - y_2)} + u_1 = 0 \tag{4.78}$$

$$f_2 - A\dot{y}_1 + a_3 S\sqrt{2g(y_2 - y_1)} - a_2 S\sqrt{2gy_1} + u_2 = 0 \tag{4.79}$$

thus, the observers for f_1 and f_2 can be built based on Eq. (4.6). For estimating fault f_1, the following ROO is proposed:

$$\dot{\gamma}_1 = -k_1(-q_{13} + u_1 + \gamma_1 + k_1 A\hat{x}_1) \tag{4.80}$$
$$\hat{f}_1 = \gamma_1 + k_1 A\hat{x}_1$$

As can be seen on Eq. (4.66), x_1 is algebraically observable, so an estimate is obtained with:

$$\dot{\gamma}_2 = -k_2(\gamma_2 + k_2 y_2)$$
$$\hat{\zeta} = \gamma_2 + k_2 y_2 \tag{4.81}$$
$$\hat{x}_1 = y_2 + \frac{1}{2ga_1^2 S^2}\left(A\hat{\zeta} + q_{32}\right)^2$$

where $\hat{\zeta}$ represents the estimation of $\dot{y}_2$.

Finally, for estimating fault f_2, the following ROO is used:

$$\dot{\gamma}_3 = -k_3(q_{32} - q_{20} + u_2 + \gamma_3 + k_3 A y_1) \tag{4.82}$$
$$\hat{f}_2 = \gamma_3 + k_3 A y_1$$

Defining $\hat{f}_1 = \hat{f}_{11}$, $\hat{f}_2 = \hat{f}_{12}$ and $\dot{\hat{f}}_2 = \hat{f}_{22}$, the chains of integrators of the faults are:

$$\dot{\hat{f}}_{11} = -k_1(-q_{13} + u_1 + \gamma_1 + k_1 A\hat{x}_1) + k_1 A\dot{\hat{x}}_1$$
$$\dot{\hat{f}}_{12} = \hat{f}_{22} \tag{4.83}$$
$$\dot{\hat{f}}_{22} = -k_3(\dot{q}_{32} - \dot{q}_{20} + \dot{u}_2 + \dot{\gamma}_3 + k_3 A\dot{y}_1) + k_3 A\ddot{y}_1$$

with

$$\dot{\hat{x}}_1 = \hat{\zeta} + \frac{A\hat{\zeta} + q_{32}}{ga_1^2 S^2}\left(A\left(-k_2(\gamma_2 + k_2 y_2) + k_2 \dot{y}_2\right) + \dot{q}_{32}\right) \tag{4.84}$$

Real-time experiments were made in the Amira DTS200 system over 3000 s, using the signals $y_{R1} = 0.06$ and $y_{R2} = 0.11$ as the references and introducing the additive faults $f_1 = 1 \times 10^{-6}(1 + sin(0.2te^{-0.01t}))\mathcal{U}(t - 220)$ and $f_2 = 1 \times 10^{-6}(1 + sin(0.05te^{-0.001t}))\mathcal{U}(t - 300)$, where $\mathcal{U}(t)$ is the step function. The design parameters were chosen as $\theta_1 = \theta_2 = 1$, $s_1 = t_1 = 1$, $s_2 = t_2 = 3$, $s_3 = t_3 = 3$, $k_1 = 1.85$, $k_2 = 0.3$, $k_3 = 22$.

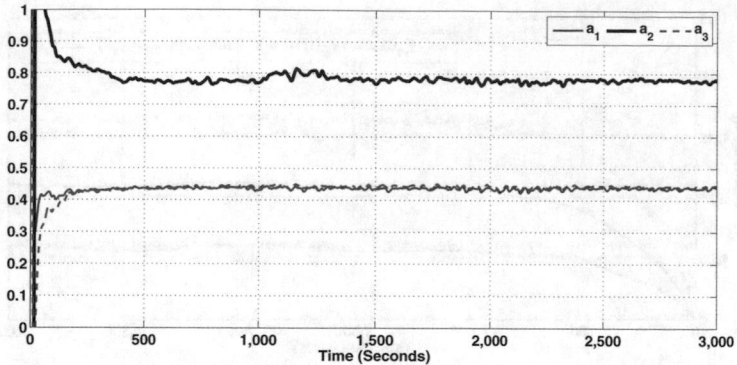

Fig. 4.7 Parameter identification of the Amira DTS200

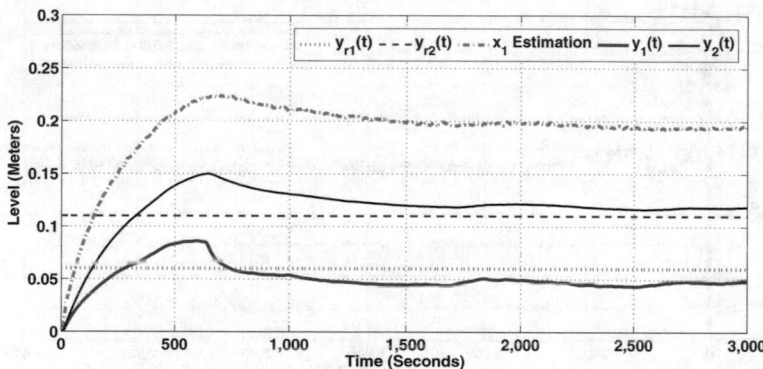

Fig. 4.8 Tank levels and multi-output tracking of the Amira DTS200 without fault-tolerant control

The parameters of the system are $A = 0.0149\,\mathrm{m}^2$ and $S = 5 \times 10^{-5}\,\mathrm{m}^2$. The unknown parameters a_1, a_2 and a_3 were estimated without the presence of faults, using the following values for the input flows: $q_1 = 0.00002\,\mathrm{m}^3/\mathrm{s}$ and $q_2 = 0.000015\,\mathrm{m}^3/\mathrm{s}$. The identification process, shown in Fig. 4.7, was performed along 3000 s, and the flow parameters were obtained as $a_1 = 0.4385$, $a_2 = 0.7774$ and $a_3 = 0.4435$.

Figures 4.8 and 4.9 compare the behavior of the tracking with and without fault-tolerance. Figure 4.8 illustrates how the effects of the faults affect tracking when a non-fault-tolerant control is applied. On the other hand, Fig. 4.9 shows how using the fault-tolerant control improves the tracking of the references, while suppressing the effects of both faults. Besides, as stated, the system operates in the desired region. The estimation $\hat{x}_1$ is also displayed.

Figures 4.10 and 4.12 show the estimation results of f_1 and f_2 respectively. These figures throw graphical valuable information, such as the magnitude of the faults and the time when they present. This information is extremely important because it can

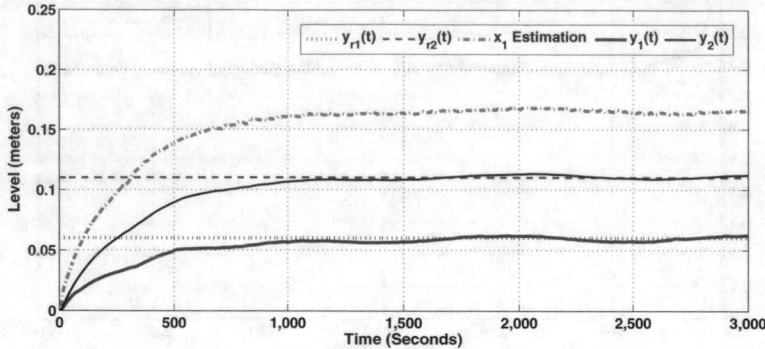

Fig. 4.9 Tank levels and multi-output tracking of the Amira DTS200 with fault-tolerant control

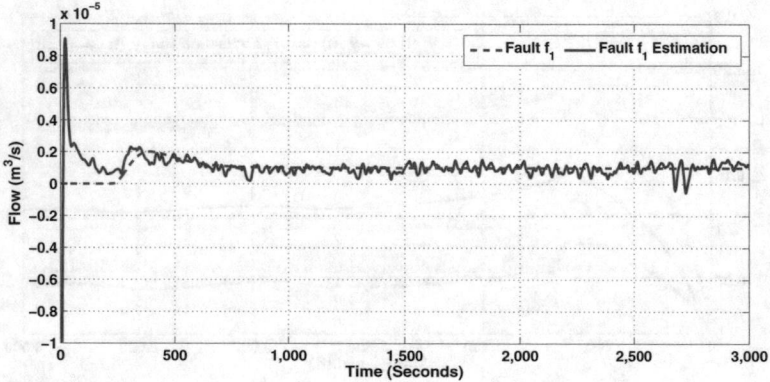

Fig. 4.10 Fault 1 estimation of the Amira DTS200

establish a critical magnitude of the fault, which may put the process on risk and forces to take corrective actions at a physical level, like sensors or actuators replacing (Fig. 4.11).

Finally, in Figs. 4.11 and 4.13 are shown the performance indices of the ROO proposed to estimate the faults f_1 and f_2. The indices were evaluated using the same cost functional used in the numerical example.

Remark 4.7 In order to reduce measurement noise, a second-order low-pass But-terworth filter was used, which has the following transfer function:

$$G_f(s) = \frac{1}{32s^2 + 8s + 1} \tag{4.85}$$

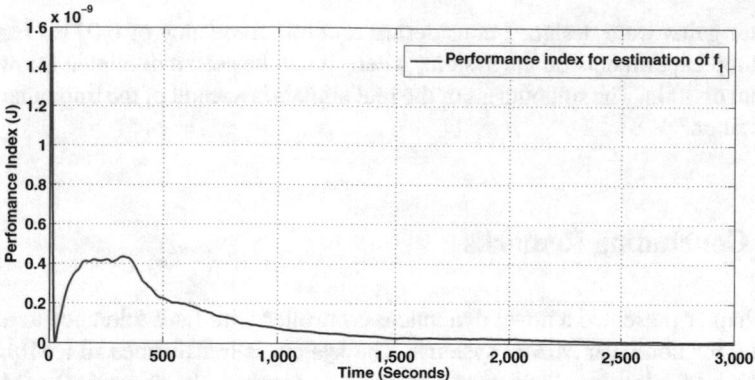

Fig. 4.11 Performance evaluation for the fault 1 estimation of the Amira DTS200

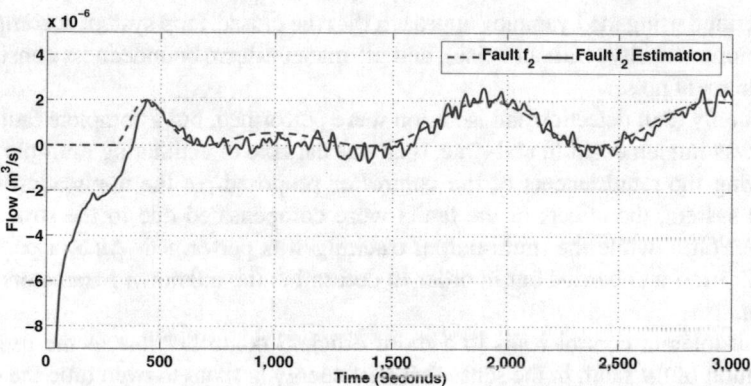

Fig. 4.12 Fault 2 estimation of the Amira DTS200

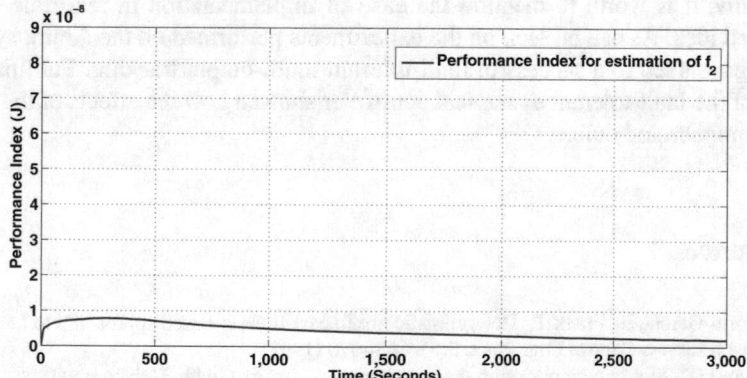

Fig. 4.13 Performance evaluation for the fault 2 estimation of the Amira DTS200

Filter gains were designed considering a cutoff frequency of 0.03 Hz based on open-loop experiments on the system, where it can be seen a dominant (slow) time constant of 300 s. The smoothness of the real signals is a result of the implementation of this filter.

4.5   Concluding Remarks

This chapter presented a novel dynamical controller with fault tolerance to achieve tracking for nonlinear MIMO systems. The system is transformed to a MMGOCF as a chain of integrators, which will be used to compute the dynamical controllers and the estimated faults. The controller is capable of suppressing the negative effects of the faults in the system, simultaneously in the presence of multiple faults. It was verified using the Lyapunov approach that the closed-loop system accomplishes asymptotic stability (without noise), and ultimate uniform boundedness considering measurement noise.

Not only fault detection and isolation were performed, but a complete fault diagnosis was implemented in real-time, that was capable of estimating multiple faults, satisfying the requirements of the controller proposed. In the application on the Amira system, the effects of the faults were compensated due to the structure of the controller, while the multi-output tracking was performed. Also, a parameter identification was carried out in order to determine the unknown parameters of the system.

Fault-tolerant control leads to a more efficient controller due to the use of an estimation of the fault, in the sense that less energy is spent to overcome the effects of the fault. The energy levels used by the control depend in a great extent on the dynamical behavior and physical nature of the fault.

Finally, it is worth to mention the ease of implementation in real-time of the proposed idea. As can be seen on the experiments performed on the Amira system, the diagnosis led to a successful fault-tolerant multi-output tracking. Fulfilling the needs of the fault-tolerant dynamical controller showed how the effects of the faults were compensated online.

References

1. Alcorta-García, E., Frank, P.: Deterministic nonlinear observer-based approaches to fault diagnosis: a survey. Control Eng. Pract. 5(5), 663–670 (1997)
2. Amira DTS200: Laboratory setup three tank system. Amira Gmbh, Duisburgh (1996)
3. Blanke, M., Kinnaert, M., Lunze, J., Staroswiecki, M.: Diagnosis and Fault-Tolerant Control. Springer, Berlin (2003)
4. Chen, L., Liu, S.: Fault diagnosis and fault-tolerant control for a nonlinear electro-hydraulic system. In: Proceedings of the Conference on Control and Fault Tolerant Systems, Nice, France, October 6–8, pp. 269–274 (2010)

5. Corless, M.J., Leitmann, G.: Continuous state feedback guaranteeing uniform ultimate bound-edness for uncertain dynamic systems. IEEE Trans. Autom. Control **26**(5), 1139–1144 (1981)
6. De Persis, C., Isidori, A.: A geometric approach to nonlinear fault detection and isolation. IEEE Trans. Autom. Control **46**(6), 853–865 (2001)
7. Fliess, M., Join, C., Sira-Ramírez, H.: Closed-Loop Fault-Tolerant Control for Uncertain Non-linear Systems. In: Meurer, T., et al. (eds.) Control and Observer Design LNCIS 322, pp. 217–233. Springer, Berlin (2005)
8. Gauthier, J.P., Hammouri, H., Othman, S.: A simple observer for nonlinear systems applications to bioreactors. IEEE Trans. Autom. Control **37**(6), 875–880 (1992)
9. Hammouri, H., Kinnaert, M., El Yaagoubi, E.H.: Observer based approach to fault detection and isolation for nonlinear systems. IEEE Trans. Autom. Control **44**(10), 1879–1884 (1999)
10. Hutter, F., Dearden, R.: Efficient on-line fault diagnosis for nonlinear systems. In: Proceedings of the 7th International Symposium on Artificial Intelligence, Robotics and Automation in Space (2003)
11. Join, C., Ponsart, J.-C., Sauter, D., Theilliol, D.: Nonlinear filter design for fault diagnosis: application to the three-tank system. IEE Proc.-Control Theory Appl. **152**(1), 55–64 (2005)
12. Join, C., Sira-Ramírez, H., Fliess, M.: Control of an uncertain three tank system via on-line parameter identification and fault detection. In: Proceedings of 16th Triennial World IFAC Conference, Prague, Czech Republic, July 2005, vol. 38, no. 1, pp. 251–256 (2005)
13. Kiltz, L., Join, C., Mboup, M., Rudolph, J.: Fault-tolerant control based on algebraic derivative estimation applied on a magnetically supported plate. Control Eng. Pract. **26**(2014), 107–115 (2014)
14. Mai, P., Hillermeier, C.: Fault-tolerant tracking control for nonlinear systems based on deriva-tive estimation. In: Proceedings of the American Control Conference, Baltimore, MD, USA, June 30–July 2, pp. 6486–6493 (2010)
15. Martínez-Guerra, R., González-Galan, R., Luviano-Juárez, A., Cruz-Victoria, J.: Diagnosis for a class of non-differentially flat and Liouvillian systems. IMA J. Math. Control I **24**(2), 177–195 (2007)
16. Martínez-Guerra, R., Mata-Machuca, J.L.: Fault Detection and Diagnosis in Nonlinear Sys-tems: A Differential and Algebraic Viewpoint. Springer, Cham (2014)
17. Martínez-Guerra, R., Mata-Machuca, J.L., Rincón-Pasaye, J.J.: Fault diagnosis viewed as a left invertibility problem. ISA Trans. **52**(5), 652–661 (2013)
18. Massoumnia, M.A., Verghese, G.C., Willsky, A.S.: Failure detection and identification. IEEE Trans. Autom. Control **34**(3), 316–321 (1989)
19. Patton, R.J.: Fault-tolerant control systems: The 1997 situation. In: IFAC Symposium on Fault Detection Supervision and Safety for Technical Processes, vol. 3, pp. 1033–1054 (1997)
20. Seron, M.M., De Don, J.A., Richter, J.H.: Integrated sensor and actuator fault-tolerant control. Int. J. Control **86**(4), 689–708 (2013)
21. Trejo-Zúñiga, I., Martínez-Guerra, R.: An invariant observer for fault diagnosis: a real-time application. In: XXI Congreso de la Asociación Chilena de Control Automático ACCA 2014, Santiago de Chile, 5–7 de Noviembre, pp. 393–398 (2014)
22. Willsky, A.: A survey of design methods in observer-based fault detection systems. Automatica **1**(2), 601–611 (1976)
23. Xu, A., Zhang, Q.: Nonlinear system fault diagnosis based on adaptive estimation. Automatica **40**(7), 1181–1193 (2004)
24. Yin, S., Ding, S.X., Xie, X., Luo, H.: A review on basic data-driven approaches for industrial process monitoring. IEEE Trans. Ind. Electron. **61**(11), 6418–6428 (2014)
25. Yin, S., Huang, Z.: Performance monitoring for vehicle suspension system via fuzzy positivistic C-means clustering based on accelerometer measurements. IEEE-ASME Trans. Mech. **20**(5), 2613–2620 (2015)

Chapter 5
Fundamentals of Fractional Calculus and Fractional Dynamical Systems

In this chapter, the most important mathematical tools related to fractional calculus are presented. Besides, it is presented an introduction to the theory of dynamical systems with dynamical equations that present fractional-order integrals and derivatives, as well as some fractional-order controllers that have been proposed, and some stability results developed for this kind of systems.

5.1 Some Functions Used in Fractional Calculus

5.1.1 Gamma Function

The gamma function $\Gamma(\cdot)$ is defined by the following integral

$$\Gamma(n) = \int_0^\infty e^{-t} t^{n-1} dt$$

which converges for real positive and noninteger negative arguments, as well as for complex arguments with positive or noninteger negative real part. Particularly, observe that $\Gamma(1) = 1$.

One of the basic properties that satisfies the gamma function is the following:

$$\Gamma(n + 1) = n\Gamma(n)$$

© The Author(s), under exclusive license to Springer Nature Switzerland AG 2021
R. Martínez-Guerra et al., *Fault-tolerant Control and Diagnosis for Integer and Fractional-order Systems*, Studies in Systems, Decision and Control 328,
https://doi.org/10.1007/978-3-030-62094-3_5

Recursively, one can obtain:

$$
\begin{aligned}
\Gamma(n+1) &= n\Gamma(n) \\
&= n(n-1)\Gamma(n-1) \\
&= n(n-1)(n-2)\Gamma(n-2) \\
&\quad \cdots \\
&= n(n-1)(n-2)\ldots\Gamma(1) \\
&= n(n-1)(n-2)\ldots 1 \\
&= n!
\end{aligned}
$$

Therefore, gamma function is a generalization of the factorial of a function, which permits to work with noninteger or even complex arguments.

5.1.2   Beta Function

The beta function $B(\cdot, \cdot)$ is defined by the following integral

$$
B(p, q) = \int_0^1 \tau^{p-1}(1-\tau)^{q-1}dt \qquad\qquad Re(p), Re(q) > 0
$$

As it can be seen, beta function is a convolution of functions t^{p-1} and t^{q-1}, when the convolution integral takes value at $t = 1$.

One of the basic properties that satisfies the beta function is the following:

$$
B(p, q) = \frac{\Gamma(p)\Gamma(q)}{\Gamma(p+q)}
$$

Beta function is used to prove some results in fractional calculus, as will be seen later.

5.1.3   Mittag-Leffler Function

Definition 5.1 Let $Re(\alpha), Re(\beta) > 0$. Function $E_{\alpha,\beta}$ defined by

$$
E_{\alpha,\beta}(z) = \sum_{k=0}^{\infty} \frac{z^k}{\Gamma(k\alpha + \beta)}
$$

whenever the series converges, is called the two-parameter Mittag-Leffler function.

If $\beta = 1$, we have the one-parameter Mittag-Leffler function

$$E_\alpha(z) = \sum_{k=0}^{\infty} \frac{z^k}{\Gamma(k\alpha + 1)}$$

Note that

$$E_1(z) = \sum_{k=0}^{\infty} \frac{z^k}{\Gamma(k + 1)}$$

$$= \sum_{k=0}^{\infty} \frac{z^k}{k!}$$

$$= e^z$$

Therefore, Mittag-Leffler function is a generalization of the exponential function, and with the right choice of the parameters α and β, it may be used to represent another kind of functions, such as the trigonometric, hyperbolic and the error function.

Moreover, consider the three-parameter Mittag-Leffler function, known also as the Prabhakar function

$$E_{\alpha,\beta}^{\gamma}(z) = \sum_{k=0}^{\infty} \frac{(\gamma)_n \, z^k}{k! \, \Gamma(k\alpha + \beta)} \qquad Re(\alpha), \, Re(\beta), \gamma > 0$$

where $(\gamma)_n = \gamma(\gamma + 1)\ldots(\gamma + n - 1)$.

Observe that if $\gamma = 1$, we recover the two-parameter Mittag-Leffler function.

5.2 Fractional-Order Integrals and Derivatives

5.2.1 Riemann–Liouville Fractional-Order Integral

Consider the integer-order antiderivative or primitive of a function $f(t)$

$$D^{-1}f(t) = I^1 \, f(t) = \int_0^t f(x)dx$$

Applying recursively the operator, we have

$$I^2 \, f(t) = \int_0^t \int_0^x f(y)dydx.$$

Changing the limits of integration, we have

$$I^2 f(t) = \int_0^t \int_y^t f(y) dx dy$$

$$= \int_0^t (t - y) f(y) dy.$$

Similarly, we obtain

$$I^3 f(t) = \frac{1}{2} \int_0^t (t - y)^2 f(y) dy$$

$$I^4 f(t) = \frac{1}{6} \int_0^t (t - y)^3 f(y) dy$$

$$I^5 f(t) = \frac{1}{24} \int_0^t (t - y)^4 f(y) dy$$

$$\vdots$$

$$I^n f(t) = \frac{1}{(n - 1)!} \int_0^t f(y)(t - y)^{n-1} dy.$$

This last equation is known as the Cauchy formula for the iterated integral. Generalizing it to the case when $n \in \mathbb{R}^+$, we have the following definition.

Definition 5.2 The fractional Riemann–Liouville integral of order $\alpha > 0$ of a function $f(t)$ is defined as

$$_0I_t^\alpha f(t) := \frac{1}{\Gamma(\alpha)} \int_0^t (t - \tau)^{\alpha-1} f(\tau) d\tau.$$

To simplify notation, the fractional Riemann–Liouville integral operator of order α will be represented as I^α.

5.2.2 Grünwald–Letnikov Fractional-Order Derivative

Consider now the definition of the integer-order derivative of a function $f(t)$

$$D^1 f(t) = \frac{df(t)}{dt} = \lim_{h \to 0} \frac{f(t) - f(t - h)}{h}$$

this is, as the limit of a backward difference. Similarly

$$D^2 f(t) = \frac{d^2 f(t)}{dt^2} = \lim_{h \to 0} \frac{1}{h^2} [f(t) - 2f(t - h) + f(t - 2h)]$$

$$\vdots$$

$$D^n f(t) = \frac{d^n f(t)}{dt^n} = \lim_{h \to 0} \frac{1}{h^n} \sum_{k=0}^{n} (-1)^n \binom{n}{k} f(t - kh)$$

where the binomial coefficient is defined as

$$\binom{n}{k} = \frac{n(n-1)(n-2)\ldots(n-k+1)}{k!}$$

If one takes $n \in \mathbb{R}^+$ in the previous equation, one obtains the fractional Grünwald–Letnikov derivative of order α

$$D^\alpha f(t) = \lim_{h \to 0} \frac{1}{h^\alpha} \sum_{k=0}^{\infty} (-1)^k \binom{\alpha}{k} f(t - kh) \qquad \forall \alpha \in \mathbb{R}^+$$

where the binomial coefficient is defined as

$$\binom{\alpha}{k} = \frac{\Gamma(\alpha + 1)}{\Gamma(k+1)\Gamma(\alpha - k + 1)}$$

This definition is useful for numerical implementations of fractional derivatives.

Moreover, considering in this definition a class of functions $f(t)$ with $m + 1$ continuous derivatives in $t > 0$, we obtain the following definition.

5.2.3   Riemann–Liouville Fractional-Order Derivative

Definition 5.3   The fractional Riemann–Liouville derivative of order α is defined as

$$_0^{RL} D_t^\alpha f(t) := D^m I^{m-\alpha} f(t) = \frac{d^m}{dt^m} \left[\frac{1}{\Gamma(m-\alpha)} \int_0^t \frac{f(\tau)}{(t-\tau)^{\alpha-m+1}} d\tau \right]$$

where $m - 1 < \alpha < m$, $m \in \mathbb{N}$.

This definition plays an important role in the development of the theory of fractional calculus and it possesses several applications in pure mathematics, such as solutions of differential equations, definitions of classes of functions, sums of series, among other.

Example 5.1   Consider the constant function $f(t) = c$. Its Riemann–Liouville fractional derivative (RLFD) is

$$^{RL}D(c) = \frac{d}{dt}\left[\frac{1}{\Gamma(1-\alpha)}\int_0^t (t-\tau)^{-\alpha}c\,d\tau\right]$$

$$= \frac{1}{\Gamma(1-\alpha)}\int_0^t \frac{d}{dt}\left[(t-\tau)^{-\alpha}c\right]d\tau$$

$$= \frac{1}{\Gamma(1-\alpha)}\int_0^t \left(-\alpha c(t-\tau)^{-\alpha-1}\right)d\tau$$

$$= -\frac{c\alpha}{\Gamma(1-\alpha)}\int_0^t (t-\tau)^{-\alpha-1}d\tau$$

$$= \frac{c\alpha}{\Gamma(1-\alpha)}\int_0^t (t-\tau)^{-\alpha-1}(-1)d\tau$$

$$= \frac{c\alpha}{\Gamma(1-\alpha)}\left[\frac{(t-\tau)^{-\alpha}}{-\alpha}\right]_0^t$$

$$= -\frac{c}{\Gamma(1-\alpha)}\left[0^{-\alpha}-(t)^{-\alpha}\right]$$

$$= \frac{c\,t^{-\alpha}}{\Gamma(1-\alpha)}.$$

Example 5.2 The RLFD of the function $f(t) = t^n$ is

$$^{RL}D(t^n) = \frac{d}{dt}\left[\frac{1}{\Gamma(1-\alpha)}\int_0^t (t-\tau)^{-\alpha}\tau^n d\tau\right]$$

Consider the following change of variables

$$\tau = ts$$
$$d\tau = t\,ds$$

Then we have

$$^{RL}D(t^n) = \frac{d}{dt}\left[\frac{1}{\Gamma(1-\alpha)}\int_0^1 (t-ts)^{-\alpha}(ts)^n t\,ds\right]$$

$$= \frac{d}{dt}\left[\frac{1}{\Gamma(1-\alpha)}\int_0^1 (t(1-s))^{-\alpha}(ts)^n t\,ds\right]$$

$$= \frac{d}{dt}\left[\frac{1}{\Gamma(1-\alpha)}\int_0^1 t^{n-\alpha+1}s^n(1-s)^{-\alpha}ds\right]$$

$$= \frac{1}{\Gamma(1-\alpha)}\int_0^1 \frac{d}{dt}\left[s^n(1-s)^{-\alpha}t^{n-\alpha+1}\right]ds$$

$$= \frac{1}{\Gamma(1-\alpha)} \int_0^1 s^n (1-s)^{-\alpha} (n-\alpha+1) t^{n-\alpha} ds$$

$$= \frac{(n-\alpha+1) t^{n-\alpha}}{\Gamma(1-\alpha)} \int_0^1 s^n (1-s)^{-\alpha} ds$$

$$= \frac{(n-\alpha+1) t^{n-\alpha}}{\Gamma(1-\alpha)} B(n+1, -\alpha+1)$$

$$= \frac{(n-\alpha+1)}{\Gamma(1-\alpha)} \frac{\Gamma(n+1)\Gamma(-\alpha+1)}{\Gamma(n-\alpha+2)} t^{n-\alpha}$$

$$= \frac{(n-\alpha+1)\Gamma(n+1)}{(n-\alpha+1)\Gamma(n-\alpha+1)} t^{n-\alpha}$$

$$= \frac{\Gamma(n+1)}{\Gamma(n-\alpha+1)} t^{n-\alpha}.$$

Example 5.3 The RLFD of the function $f(t) = e^{nt}$ is

$$^{RL}D(e^{nt}) = \sum_{k=0}^{\infty} \frac{n^k}{k!} \frac{1}{\Gamma(1-\alpha)} \int_0^t \frac{d}{dt} \left[\tau^k (t-\tau)^{-\alpha} \right] d\tau$$

$$= \sum_{k=0}^{\infty} \frac{n^k}{k!} \frac{-\alpha}{\Gamma(1-\alpha)} \int_0^t \tau^k (t-\tau)^{-\alpha-1} d\tau$$

Consider the following change of variables

$$\tau = ts$$

$$d\tau = tds$$

Then we have

$$^{RL}D(e^{nt}) = \sum_{k=0}^{\infty} \frac{n^k}{k!} \frac{-\alpha}{\Gamma(1-\alpha)} \int_0^1 (ts)^k (t(1-s))^{-\alpha-1} tds$$

$$= \sum_{k=0}^{\infty} \frac{n^k}{k!} \frac{-\alpha}{\Gamma(1-\alpha)} t^k t^{-\alpha-1} t \int_0^1 s^k (1-s)^{-\alpha-1} ds$$

$$= \sum_{k=0}^{\infty} \frac{n^k}{k!} \frac{-\alpha}{\Gamma(1-\alpha)} t^{k-\alpha} B(k+1, -\alpha)$$

$$= \sum_{k=0}^{\infty} \frac{n^k}{k!} \frac{-\alpha}{\Gamma(1-\alpha)} \frac{\Gamma(k+1)\Gamma(-\alpha)}{\Gamma(k+1-\alpha)} t^{k-\alpha}$$

$$= \sum_{k=0}^{\infty} \frac{n^k}{\Gamma(k+1)} \frac{-\alpha}{\Gamma(1-\alpha)} \frac{\Gamma(k+1)\Gamma(-\alpha)}{\Gamma(k+1-\alpha)} t^{k-\alpha}$$

$$= \sum_{k=0}^{\infty} \frac{n^k}{\Gamma(k+1)} \frac{-\alpha}{-\alpha\Gamma(-\alpha)} \frac{\Gamma(k+1)\Gamma(-\alpha)}{\Gamma(k+1-\alpha)} t^{k-\alpha}$$

$$= \sum_{k=0}^{\infty} \frac{n^k}{\Gamma(k+1-\alpha)} t^{k-\alpha}$$

$$= t^{-\alpha} \sum_{k=0}^{\infty} \frac{(nt)^k}{\Gamma(k+1-\alpha)}$$

$$= t^{-\alpha} E_{1,1-\alpha}(nt).$$

5.2.4   Caputo Fractional-Order Derivative

Some application problems, such as those existing in viscoelasticity theory and solid mechanics, require the formulation of initial conditions for their modelling. However, the Riemann–Liouville derivative leads to initial conditions that include limit values of fractional derivatives; even if these problems may be solved mathematically, the solutions thus obtained are practically useless since there is not physical meaning for these initial conditions. The solution proposed by M. Caputo for this problem is the following.

Definition 5.4 The Caputo fractional derivative is defined as

$$_0^C D_t^\alpha f(t) := I^{m-\alpha} D^m f(t) = \frac{1}{\Gamma(m-\alpha)} \int_0^t \frac{f^{(m)}(\tau)}{(t-\tau)^{\alpha-m+1}} d\tau$$

where $m - 1 < \alpha < m$, $m \in \mathbb{N}$.

The principal advantage of Caputo's approach is that the initial conditions of differential equations with derivatives of this kind take the same form as in the integer case, which possess known physical meaning. Nevertheless, it is worth noting that Caputo's definition requires the function f to be at least m times differentiable.

Given that this text deals with applications to models of physical systems, the only definition of fractional derivative used will be the one from Caputo, whose operator of order α will be denoted as D^α to simplify notation.

Example 5.4 Consider the constant function $f(t) = c$. Its Caputo fractional derivative (CFD) is

$$D(c) = \frac{1}{\Gamma(1-\alpha)} \int_0^t (t-\tau)^{-\alpha} \frac{d}{d\tau}(c)\, d\tau$$

$$= \frac{1}{\Gamma(1-\alpha)} \int_0^t (t-\tau)^{-\alpha}(0)\, d\tau$$

$$= 0.$$

This is another advantage of the CFD with respect to the RLFD. The result of 0 for the derivative of a constant is a more natural result, expected for physical applications.

Example 5.5 The CFD of the function $f(t) = t^n$ is

$$D(t^n) = \frac{1}{\Gamma(1-\alpha)} \int_0^t (t-\tau)^{-\alpha} \frac{d}{d\tau}(\tau^n)\, d\tau$$

$$= \frac{n}{\Gamma(1-\alpha)} \int_0^t (t-\tau)^{-\alpha}\tau^{n-1}\, d\tau$$

$$= \frac{\Gamma(n+1)}{\Gamma(n)\Gamma(1-\alpha)} \int_0^t (t-\tau)^{-\alpha}\tau^{n-1}\, d\tau$$

Consider the following change of variables

$$\tau = ts$$
$$d\tau = t\, ds$$

Then we have

$$D(t^n) = \frac{\Gamma(n+1)}{\Gamma(n)\Gamma(1-\alpha)} \int_0^t (t-\tau)^{-\alpha}\tau^{n-1}\, d\tau$$

$$= \frac{\Gamma(n+1)}{\Gamma(n)\Gamma(1-\alpha)} \int_0^1 (t(1-s))^{-\alpha}(ts)^{n-1} t\, ds$$

$$= \frac{\Gamma(n+1)}{\Gamma(n)\Gamma(1-\alpha)} t^{n-\alpha} \int_0^1 s^{n-1}(1-s)^{-\alpha}\, ds$$

$$= \frac{\Gamma(n+1)}{\Gamma(n)\Gamma(1-\alpha)} t^{n-\alpha} B(n, -\alpha+1)$$

$$= \frac{\Gamma(n+1)}{\Gamma(n)\Gamma(1-\alpha)} \frac{\Gamma(n)\Gamma(-\alpha+1)}{\Gamma(n-\alpha+1)} t^{n-\alpha}$$

$$= \frac{\Gamma(n+1)}{\Gamma(n-\alpha+1)} t^{n-\alpha}.$$

Example 5.6 The CFD of the function $f(t) = e^{nt}$ is

$$D\left(e^{nt}\right) = \frac{1}{\Gamma(1-\alpha)} \int_0^t (t-\tau)^{-\alpha} \frac{d}{d\tau}\left[e^{n\tau}\right] d\tau$$

$$= \frac{1}{\Gamma(1-\alpha)} \int_0^t (t-\tau)^{-\alpha} \frac{d}{d\tau}\left[\sum_{k=0}^{\infty} \frac{(n\tau)^k}{k!}\right] d\tau$$

$$= \sum_{k=0}^{\infty} \frac{n^k}{k!} \frac{1}{\Gamma(1-\alpha)} \int_0^t (t-\tau)^{-\alpha} \frac{d}{d\tau}\left[\tau^k\right] d\tau$$

$$= \sum_{k=0}^{\infty} \frac{n^k}{k!} \frac{1}{\Gamma(1-\alpha)} \int_0^t (t-\tau)^{-\alpha} k\tau^{k-1} d\tau$$

Consider the following change of variables

$$\tau = ts$$
$$d\tau = t\,ds$$

Then we have

$$D\left(e^{nt}\right) = \sum_{k=0}^{\infty} \frac{n^k}{k!} \frac{k}{\Gamma(1-\alpha)} \int_0^1 (ts)^{k-1} (t(1-s))^{-\alpha} t\,ds$$

$$= \sum_{k=0}^{\infty} \frac{n^k}{k!} \frac{k}{\Gamma(1-\alpha)} t^{k-1} t^{-\alpha} t \int_0^1 s^{k-1} (1-s)^{-\alpha} ds$$

$$= \sum_{k=0}^{\infty} \frac{n^k}{k!} \frac{k}{\Gamma(1-\alpha)} t^{k-\alpha} B(k, -\alpha + 1)$$

$$= \sum_{k=0}^{\infty} \frac{n^k}{k!} \frac{k}{\Gamma(1-\alpha)} \frac{\Gamma(k)\Gamma(-\alpha+1)}{\Gamma(k-\alpha+1)} t^{k-\alpha}$$

$$= \sum_{k=0}^{\infty} \frac{n^k}{(k-1)!} \frac{1}{\Gamma(1-\alpha)} \frac{\Gamma(k)\Gamma(-\alpha+1)}{\Gamma(k-\alpha+1)} t^{k-\alpha}$$

$$= \sum_{k=0}^{\infty} \frac{n^k}{\Gamma(k)} \frac{1}{\Gamma(1-\alpha)} \frac{\Gamma(k)\Gamma(-\alpha+1)}{\Gamma(k-\alpha+1)} t^{k-\alpha}$$

$$= \sum_{k=0}^{\infty} \frac{n^k}{\Gamma(k - \alpha + 1)} t^{k-\alpha}$$

$$= t^{-\alpha} \sum_{k=0}^{\infty} \frac{(nt)^k}{\Gamma(k - \alpha + 1)}$$

$$= t^{-\alpha} E_{1,1-\alpha}(nt).$$

In this two last examples, the results obtained were the same as the ones obtained with the RLFD. However, this must not be expected in general, as there is a difference in the definition of the operators.

5.2.5 Some Considerations for Fractional-Order Operators

5.2.5.1 Difference Between the RLFD and CFD Operators

Consider the expansion of the function $f(t)$ in its polynomial form, from Taylor's theorem

$$f(t) = \sum_{k=0}^{n-1} \frac{t^k}{k!} f^{(k)}(0) + \frac{1}{(n-1)!} \int_0^t (t - \tau)^{n-1} f^{(n)}(\tau) d\tau$$

$$= \sum_{k=0}^{n-1} \frac{t^k}{k!} f^{(k)}(0) + I^n f^{(n)}$$

where $I^n f^{(n)}$ is the integral form of the remainder. Applying the RLFD to the equation, we have

$$^{RL}D^{\alpha} f(t) = {}^{RL}D^{\alpha} \left[\sum_{k=0}^{n-1} \frac{t^k}{k!} f^{(k)}(0) + I^n f^{(n)} \right]$$

$$= \sum_{k=0}^{n-1} \frac{{}^{RL}D^{\alpha}(t^k)}{\Gamma(k+1)} f^{(k)}(0) + {}^{RL}D^{\alpha} I^n D^n(f)$$

Given that $^{RL}D^{\alpha}(t^k) = \frac{\Gamma(k+1)}{\Gamma(k-\alpha+1)} t^{k-\alpha}$, we have

$$^{RL}D^\alpha f(t) = \sum_{k=0}^{n-1} \frac{\Gamma(k+1)t^{k-\alpha}}{\Gamma(k-\alpha+1)\Gamma(k+1)} f^{(k)}(0) + D^n I^{n-\alpha} I^n D^n(f)$$

$$= \sum_{k=0}^{n-1} \frac{t^{k-\alpha}}{\Gamma(k-\alpha+1)} f^{(k)}(0) + D^n I^n I^{n-\alpha} D^n(f)$$

$$= \sum_{k=0}^{n-1} \frac{t^{k-\alpha}}{\Gamma(k-\alpha+1)} f^{(k)}(0) + I^{n-\alpha} D^n(f)$$

$$= \sum_{k=0}^{n-1} \frac{t^{k-\alpha}}{\Gamma(k-\alpha+1)} f^{(k)}(0) + {}^C D^\alpha f(t)$$

In this relation can be seen the additional terms that make both differential operators to differ. In a similar manner, one can get the following relation

$$^C D^\alpha f(t) = {}^{RL}D^\alpha \left(f(t) - \sum_{k=0}^{m-1} f^{(k)}(0^+) \frac{t^k}{\Gamma(k+1)} \right).$$

5.2.5.2 Fundamental Theorem of Calculus

Recall the Fundamental Theorem of Calculus for the integer-order case.
   (a) First part. Let $F(x)$ be an antiderivative of $f(x)$. Then

$$F(x) = \int_a^x f(t)dt \quad \Longrightarrow \quad F'(x) = f(x) \quad \forall x \in (a,b)$$

This roughly implies that

$$F'(x) = D^1 I^1 f(x) = f(x)$$

This means, the integer-order derivative is a left-inverse operator for the integer-order integral, and both operators nullify each other, leaving only the original function.
   (b) Second part.

$$F'(x) = f(x) \quad \Longrightarrow \quad \int_a^b f(x)dx = F(b) - F(a) \quad \forall x \in [a,b]$$

This roughly implies that

$$I^1 f(x) = I^1 D^1 F(x) = F(x)$$

This means, the integer-order derivative is also a right-inverse operator for the integer-order integral, and both operators nullify each other, leaving only the original function.

However, for the RLFD operator, we have

$$
\begin{aligned}
{}^{RL}D^{\alpha} I^{\alpha} f(t) &= f(t) \\
&\neq I^{\alpha}\, {}^{RL}D^{\alpha} f(t) \\
&= f(t) - \frac{t^{\alpha-1}}{\Gamma(\alpha)} I^{1-\alpha} f(0^+)
\end{aligned}
$$

and for the CFD operator

$$
\begin{aligned}
D^{\alpha} I^{\alpha} f(t) &= f(t) \\
&\neq I^{\alpha} D^{\alpha} f(t) \\
&= f(t) - \sum_{k=0}^{n-1} \frac{t^k}{k!} f^{(k)}(0^+)
\end{aligned}
$$

This means, both fractional-order derivatives are only left-inverse operators for the fractional-order Riemann–Liouville integral, leaving only the original function; nevertheless, as it has been stated, this is not true when applying them from the right.

5.2.5.3 Leibniz Rule

Recall now the Leibniz rule for the derivative of a product of two functions

$$
D(fg) = fg' + gf'
$$

For the nth derivative, we have

$$
D^n(fg) = \sum_{k=0}^{n} \binom{n}{k} D^{n-k}(f) D^k(g)
$$

where the binomial coefficient is defined as

$$
\binom{n}{k} = \frac{n!}{k!(n-k)!}
$$

Moreover, the Lebniz formula for the RLFD operator is

$$
{}^{RL}D^{\alpha}(fg) = \sum_{k=0}^{\lfloor \alpha \rfloor} \binom{\alpha}{k} D^k(f) D^{\alpha-k}(g) + \sum_{k=\lfloor \alpha \rfloor+1}^{\infty} \binom{\alpha}{k} D^k(f) I^{k-\alpha}(g)
$$

and for the CFD is

$$D^\alpha(fg) = \frac{t^{-\alpha}}{\Gamma(1-\alpha)}g(0)(f(t) - f(0)) + D^\alpha(g)f + \sum_{k=1}^{\infty}\binom{\alpha}{k}I^{k-\alpha}(g)D^k(f)$$

where the binomial coefficient is defined as

$$\binom{\alpha}{k} = \frac{\Gamma(\alpha+1)}{\Gamma(k+1)\Gamma(\alpha-k+1)}.$$

The problem with these definitions, as it can be seen, is that both contain an infinite sum of terms, which complicates their use in real-world applications.

5.3 Fractional-Order Differential Equations

In this section, some linear ordinary differential equations of fractional-order will be solve by means of the Laplace transform method.

5.3.1 Laplace Transform of Fractional-Order Functions

Recall the formula to obtain the Laplace transform (LT) of a function $f(t)$

$$\mathscr{L} = \int_0^\infty f(t)e^{-st}dt$$

which is used to convert the function from the t domain to the s domain.

Example 5.7 The LT of $f(t) = e^{nt}$ is

$$\mathscr{L}[e^{nt}] = \int_0^\infty e^{nt}e^{-st}dt$$

$$= \int_0^\infty e^{(n-s)t}dt$$

$$= \int_0^\infty e^{-(s-n)t}dt$$

$$= \left[-\frac{1}{s-n}e^{-(s-n)t}\right]_0^\infty$$

$$= -\frac{1}{s-n}\left[e^{-\infty} - e^0\right]$$

$$= -\frac{1}{s-n}[0-1]$$

$$= \frac{1}{s-n}.$$

Example 5.8 The LT of $f(t) = E_{\alpha,\beta}(nt^{\alpha})$ is

$$\mathscr{L}\left[E_{\alpha,\beta}(nt^{\alpha})\right] = \int_0^{\infty} E_{\alpha,\beta}(nt^{\alpha})e^{-st}\,dt$$

$$= \int_0^{\infty} \sum_{k=0}^{\infty} \frac{(nt^{\alpha})^k}{\Gamma(\alpha k+\beta)}e^{-st}\,dt$$

$$= \sum_{k=0}^{\infty} \frac{n^k}{\Gamma(\alpha k+\beta)} \int_0^{\infty} t^{\alpha k}e^{-st}\,dt$$

$$= t^{\beta-1}\sum_{k=0}^{\infty} \frac{n^k}{\Gamma(\alpha k+\beta)} \int_0^{\infty} t^{\alpha k}e^{-st}\,dt$$

$$\mathscr{L}\left[t^{\beta-1}E_{\alpha,\beta}(nt^{\alpha})\right] = \sum_{k=0}^{\infty} \frac{n^k}{\Gamma(\alpha k+\beta)} \int_0^{\infty} t^{\alpha k+\beta-1}e^{-st}\,dt$$

Consider the following change of variables

$$x = st$$
$$dx = s\,dt$$

Then we have

$$\mathscr{L}\left[t^{\beta-1}E_{\alpha,\beta}(nt^{\alpha})\right] = \sum_{k=0}^{\infty} \frac{n^k}{\Gamma(\alpha k+\beta)} \int_0^{\infty} \left(\frac{x}{s}\right)^{\alpha k+\beta-1} e^{-x}\frac{1}{s}\,dx$$

$$= \sum_{k=0}^{\infty} \frac{n^k}{\Gamma(\alpha k+\beta)} \frac{1}{s^{\alpha k+\beta}} \int_0^{\infty} x^{\alpha k+\beta-1}e^{-x}\,dx$$

$$= \sum_{k=0}^{\infty} \frac{n^k}{\Gamma(\alpha k+\beta)} \frac{1}{s^{\alpha k+\beta}}\Gamma(\alpha k+\beta)$$

$$= \sum_{k=0}^{\infty} n^k \frac{1}{s^{\alpha k+\beta}}$$

$$= s^{-\beta} \sum_{k=0}^{\infty} (ns^{-\alpha})^k$$

$$= \frac{s^{-\beta}}{1 - ns^{-\alpha}}$$

$$= \frac{s^{\alpha-\beta}}{s^{\alpha} - n}.$$

Appendix A.1 contains tables with LT for some functions. Observe the similarities between the LT of the integer and the fractional-order functions, remembering that the Mittag-Leffler function is a generalization of the exponential function. Moreover, the LT of the differential and integral operators is shown, as well as the LT of the convolution between two time functions.

5.3.2 Solution of FODE by Means of the Laplace Transform

In this section, a couple of linear fractional-order ordinary differential equations (FODE) will be solve by means of the Laplace transform, with the aid of the formulas given in the Tables of the Appendix for this Chapter.

Example 5.9 Consider the following linear homogeneous FODE

$$_{0}^{RL}D_t^{1/2} f(t) + a f(t) = 0 \qquad\qquad _0 D_t^{-1/2} f(0) = C$$

Observe the fractional-order initial conditions required for this problem. Applying the LT to this equation, and considering $m = 1$, we have

$$s^{1/2} F(s) - f^{(-1/2)}(0) + a F(s) = 0$$
$$s^{1/2} F(s) - C + a F(s) = 0$$
$$(s^{1/2} + a) F(s) = C$$
$$F(s) = \frac{C}{s^{1/2} + a}$$

Finally, applying inverse LT, we have

$$f(t) = C t^{-1/2} E_{\frac{1}{2}, \frac{1}{2}}(-a t^{1/2}).$$

Example 5.10 Consider the following linear nonhomogeneous FODE

$$_{0}^{RL}D_t^{Q} f(t) +_{0}^{RL} D_t^{q} f(t) = h(t)$$

Applying the LT to this equation, and considering $0 < q < Q < 1$ (i.e. $m = 1$), we have

$$s^Q F(s) - f^{Q-1}(0) + s^q F(s) - f^{q-1}(0) = H(s)$$
$$(s^Q + s^q)F(s) = C + H(s)$$

where

$$C = f^{Q-1}(0) + f^{q-1}(0)$$

Hence

$$F(s) = \frac{C + H(s)}{s^Q + s^q}$$

In order to find a suitable LT for this function, the following arrangement is done

$$F(s) = \frac{C + H(s)}{s^q(s^{Q-q} + 1)}$$
$$= \frac{Cs^{-q}}{s^{Q-q} + 1} + H(s)\frac{s^{-q}}{s^{Q-q} + 1}$$

Consider first the left part. Applying LT we have

$$\mathscr{L}\left[\frac{Cs^{-q}}{s^{Q-q} + 1}\right] = Ct^{Q-1}E_{Q-q,Q}\left(-t^{Q-q}\right)$$

For the right part, note that it comprises a convolution

$$\mathscr{L}\left[H(s)\frac{s^{-q}}{s^{Q-q} + 1}\right] = \mathscr{L}^{-1}[H(s)] * \mathscr{L}^{-1}\left[\frac{s^{-q}}{s^{Q-q} + 1}\right]$$
$$= h(t) * t^{Q-1}E_{Q-q,Q}\left(-t^{Q-q}\right)$$

Therefore

$$f(t) = CG(t) + \int_0^t G(t - \tau)h(\tau)d\tau$$

where

$$G(t) = t^{Q-1}E_{Q-q,Q}\left(-t^{Q-q}\right).$$

5.4 Fractional Dynamical Systems

As it has been mentioned, fractional-order dynamical systems, in contrast with the integer-order ones, have been studied strongly in the last decades. This is due to the great amount of applications and physical phenomena whose dynamics present fractional derivatives and integrals, such as diffusion problems, viscoelasticity, poly-meric behaviour, financial systems, biological systems, damped mechanical systems, electric circuits, electrochemistry, rheology, fractals and heat propagation.

Particularly, in control theory one of the most important contributions has been the development of generalized PID controllers, as well as other fractional-order controllers such as the CRONE and the fractional-order sliding mode controller. Nowadays almost every kind of controller has been extended to its fractional coun-terpart.

5.4.1 Commensurate-Order Fractional Systems

There exist different definitions for these kind of systems, e.g. the following, which is found in [1].

Definition 5.5 The fractional differential equation

$$g(x, y(x), D_{*0}^{n_1} y(x), D_{*0}^{n_2} y(x), \ldots, D_{*0}^{n_k} y(x)) = 0$$

with $0 < n_1 < n_2 < \cdots < n_k$ and a certain function g is called commensurate if the numbers $n_1, n_2, \ldots, n_k$ are commensurate, i.e. if the quotients n_μ / n_ν are rational numbers for all $\mu, \nu \in \{1, 2, \ldots, k\}$.

In this case, the author uses this definition because it is related to the traditional use of the concept common in number theory. However, in this book the following definition will be used, from [7].

Definition 5.6 Consider the following model in state space:

$$D^\alpha x = Ax + Bu$$
$$y = Cx$$

where $x \in \mathbb{R}^n$, $u \in \mathbb{R}^r$ and $y \in \mathbb{R}^p$, $\boldsymbol{\alpha} = [\alpha_1, \alpha_2, \ldots, \alpha_n]^T$ is the vector of fractional orders. If $\alpha_1 = \alpha_2 = \cdots = \alpha_n = \alpha \in \mathbb{R}$, system is called commensurate-order, oth-erwise it is an incommensurate-order system.

For this text purposes, consider the following class of commensurate-order frac-tional nonlinear systems with unknown inputs (faults):

$$D^\alpha x = g(x, u, f) \tag{5.1}$$
$$y = h(x, u)$$

where $x \in \mathbb{R}^n$ is the state vector, $u \in \mathbb{R}^m$ is the input (control) vector, $f \in \mathbb{R}^q$ is the unknown input (fault) vector, $y \in \mathbb{R}^p$ is the output vector, $\boldsymbol{\alpha} = (\alpha_1, \ldots, \alpha_n)$, g and h are analytic functions. Particularly, in this book $0 < \alpha < 1$ will be used.

5.4.2 Incommensurate-Order Fractional Systems

As stated in the past subsection, if a fractional-order system is not of commensurate-order, it is called an incommensurate-order system. Consider the following definition.

Definition 5.7 Consider the following model:

$$D^{\alpha_i} x_i = f_i (x_1, \ldots, x_n)$$

where $x \in \mathbb{R}^n$, $1 \leq i \leq n$, $i \in \mathbb{Z}^+$. If $\alpha_i \neq \alpha_j$ for at least one value of i, system is called incommensurate-order.

5.5 Fractional-Order Controllers

Fractional dynamical systems had been studied in the context of control systems in a measured way, due to the lack of appropriate mathematical methods for their analysis. In general time-domain had been avoided, and thus the first fractional-order controllers were developed in the frequency-domain [8, 9]. Between these ones, the most important are the fractional-order PID controller and the CRONE, which will be briefly described in the next sections.

Furthermore, it is worth to note that fractional calculus has also been extended to other kind of control strategies, such as the $\mathcal{H}_2$ and $\mathcal{H}_\infty$ control, lag-lead compensation, sliding mode control, model reference adaptive control, reset control and state observers [5].

5.5.1 Fractional-Order PID Controller

I. Podlubny proposed a generalization of the PID controller, known as $PI^\lambda D^\mu$ controller which, as its name implies, involves an λ-order integral term and a μ-order derivative term, where $\lambda, \mu \in \mathbb{R}$.

As it is stated in the classic (integer-order) PID controller theory, the integral term makes slower the system response and reduces its relative stability, but eliminates the steady-state error. In the time-domain, it diminishes the rise time of the transient response, while increasing the settling time and the overshoot. In the complex plane, the integral action displaces the root-locus of the system towards the right half-plane. Finally in the frequency-domain, it creates an increment of -20 dB/dec in the slopes of the magnitude curve and a constant lag of $-\pi/2$ rad to the phase curve.

Moreover, the derivative term increases system's stability, but magnifies the high-frequency noise effects. In time-domain, it can be seen a decrement in the overshoot and the settling time. In the complex plane, the derivative action displaces the root-locus of the system towards the left half-plane. Finally in the frequency-domain, it creates an increment of 20 dB/dec in the slopes of the magnitude curve and a constant lead of $\pi/2$ rad to the phase curve.

As it can be seen, in the integer case the global effects of the integral and derivative actions depend on the chosen gains, but have in general the same kind of responses. In the Laplace-domain, these actions may be seen as variables with an inverse but unit exponent, being 0 the proportional case. Hence if the values of these actions are chosen as non-integer, it is expected to have intermediate effects to the ones obtained in the classic case.

Therefore, the objective of the $PI^{\lambda}D^{\mu}$ controller is to obtain a desired response of the plant, not only by tuning the gains, but also by properly varying the orders of the integral and derivative actions. It has been proven that by means of this technique better results are obtained than with the classic PID.

5.5.2 CRONE

CRONE is the french acronym for non-integer order robust control. It was proposed by A. Oustaloup [6, 10] and represents the first theoretical frame for the application of fractional-order systems in automatic control. Some of the main features of this technique are:

- Methodology based on the frequency-domain.
- Continuous or discrete control of multiple input and output systems with disturbances.
- Use of unit feedback.
- Robustness with respect to parametric uncertainties.
- Control of minimum and non-minimum phase systems, unstable plants or with mechanical flexion modes, time-varying and nonlinear plants.

There exist three generations of CRONE controllers. The first one is used when the plant to control has a constant phase around a frequency of interest, and it makes the loop robust with respect to changes in the plant gain; however, it does not assure an asymptotic behaviour. Its transfer function is given by

$$C(s) = C_0 s^{\alpha}$$

with $\alpha, C_0 \in \mathbb{R}$.

If the plant does not have a constant phase, the second generation CRONE is used, which has the following transfer function:

$$C(s) = \frac{F(s)}{G(s)} \qquad F(s) = \left(\frac{\omega_{cg}}{s}\right)^{\alpha}$$

where ω_{cg} is the open-loop cross frequency and $\alpha \in [1, 2]$.

Finally, the third generation CRONE considers other kind of uncertainties in the model, such as the wrong assignment of zeros and poles. Its main objective is to assure that the closed-loop gain (or even the damping factor) never exceeds certain value, even when some plant parameters vary in a certain range. It has been demonstrated that also with this technique better results are achieved than with the classic PID controller.

5.6 Stability Results for Fractional-Order Systems

In this section, some existing stability results for fractional-order dynamical systems are briefly presented.

5.6.1 Commensurate-Order Systems

5.6.1.1 Linear Systems

Consider the following class of linear systems:

$$D^\alpha x = Ax + Bu, \qquad x(0) = x_0 \qquad (5.2)$$
$$y = Cx$$

Theorem 5.1 ([4]) *The autonomous system (5.2) is:*

- *asymptotically stable if and only if* $|\arg(\lambda(A))| > \alpha\pi/2$. *In this case, the components of the state decay towards 0 like* $t^{-\alpha}$.
- *stable if and only if either it is asymptotically stable, or those critical eigenvalues which satisfy* $|\arg(\lambda(A))| = \alpha\pi/2$ *have geometric multiplicity 1 (Fig. 5.1).*

5.6.1.2 Nonlinear Systems

Consider the following class of nonlinear systems:

$$D^\alpha x = f(t, x), \qquad \alpha \in (0, 1] \qquad (5.3)$$

In what follows, some results are presented, regarding stability proofs for the class of systems (5.3).

Definition 5.8 The solution of system (5.3) is said to be Mittag-Leffler stable if

$$\|x\| \le \{m[x(0)]E_{\alpha,1}(-\lambda t^\alpha)\}^b$$

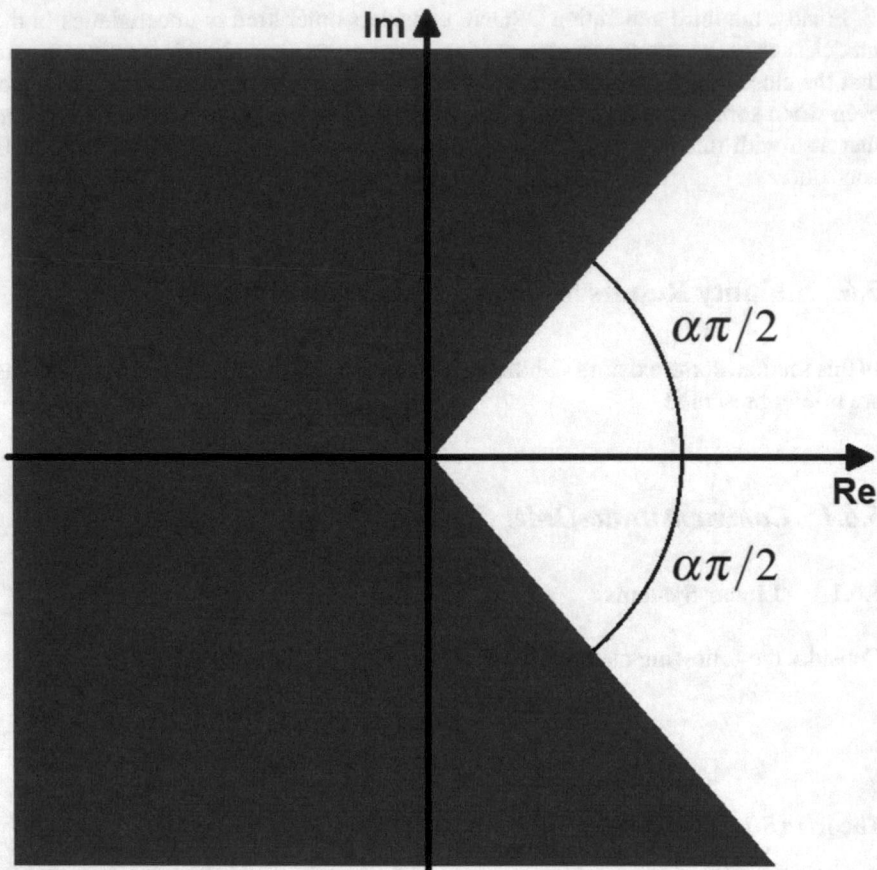

Fig. 5.1 Stability region for linear autonomous fractional-order systems

$\alpha \in (0, 1), \lambda \geq 0, b > 0, m(0) = 0, m(x) \geq 0$, and $m(x)$ is locally Lipschitz (with Lipschitz constant m_0) on $x \in \mathbb{B}$, which is an open subset of $\mathbb{R}^n$.

Theorem 5.2 ([3]) *Let $x = 0$ be an equilibrium point for the system (5.3) and let $\mathbb{D} \subset \mathbb{R}^n$ be a domain containing the origin. Let $V(t, x) : [0, \infty) \times \mathbb{D} \to \mathbb{R}$ be a continuously differentiable function and locally Lipschitz with respect to x such that*

$$\alpha_1 \|x\|^a \leq V(t, x) \leq \alpha_2 \|x\|^{ab}$$
$$D^\beta V(t, x) \leq -\alpha_3 \|x\|^{ab}$$

where $t \geq 0, x \in \mathbb{D}, \beta \in (0, 1), \alpha_1, \alpha_2, \alpha_3, a$ and b are arbitrary positive constants. Then $x = 0$ is Mittag-Leffler stable. If the assumptions hold globally on $\mathbb{R}^n$, then $x = 0$ is globally Mittag-Leffler stable.

Lemma 5.1 ([2]) *Let $x \in \mathbb{R}^n$ be a vector of differentiable functions. Then, for any time instant $t \geq t_0$, the following relationship holds*

$$\frac{1}{2} D^\alpha (x^T P x) \leq x^T P D^\alpha x, \qquad \forall \alpha \in (0, 1], \forall t \geq t_0$$

where $P \in \mathbb{R}^{n \times n}$ is a constant, symmetric and positive definite matrix.

5.6.2   Incommensurate-Order Systems

Consider the following class of linear incommensurate-order systems:

$$D^{\alpha_1} x_1 = a_{11} x_1 + a_{12} x_2 + \cdots + a_{1n} x_n \qquad (5.4)$$
$$D^{\alpha_2} x_2 = a_{21} x_1 + a_{22} x_2 + \cdots + a_{2n} x_n$$

$$\vdots$$

$$D^{\alpha_n} x_n = a_{n1} x_1 + a_{n2} x_2 + \cdots + a_{nn} x_n$$

where α_i is a rational number between 0 and 1.

Let $\alpha_i = v_i / u_i$, $(u_i, v_i) = 1$, $u_i, v_i \in \mathbb{Z}^+$, $i = 1, \ldots, n$, Let M be the lowest common multiple of the denominators u_i, and set $\gamma = 1/M$. Then the zero solution of system (5.4) is Lyapunov globally asymptotically stable if all the roots $_i$ of the equation

$$\det \begin{bmatrix} \lambda_1^{M\alpha_1} - a_{11} & -a_{12} & \cdots & -a_{1n} \\ -a_{21} & \lambda_2^{M\alpha_2} - a_{22} & \cdots & -a_{2n} \\ \vdots & \vdots & \ddots & \vdots \\ -a_{n1} & -a_{n2} & \cdots & \lambda_n^{M\alpha_n} - a_{11} - a_{nn} \end{bmatrix} = 0$$

satisfy $|arg(\lambda_i)| > \gamma \alpha / 2$.

References

1. Diethelm, K.: The Analysis of Fractional Differential Equations: An Application-Oriented Exposition Using Differential Operators of Caputo Type. Springer, Berlin (2010)
2. Duarte-Mermoud, M.A., Aguila-Camacho, N., Gallegos, J.A., Castro-Linares, R.: Using general quadratic Lyapunov functions to prove Lyapunov uniform stability for fractional order systems. Commun. Nonlinear Sci. **22**(1–3), 650–659 (2015)
3. Li, Y., Chen, Y., Podlubny, I.: Stability of fractional-order nonlinear dynamic systems: Lyapunov direct method and generalized Mittag-Leffler stability. Comput. Math. Appl. **59**(5), 1810–1821 (2010)

4. Matignon, D.: Stability results for fractional differential equations with applications to control processing. Ma Comput. Sci. Eng. **2**(1996), 963–968 (1996)
5. Monje, C.A., Chen, Y., Vinagre, B.M., Xue, D., Feliu, V.: Fractional-Order Systems and Controls: Fundamentals and Applications. Springer, London (2010)
6. Oustaloup, A.: La commande CRONE: commande robuste d'ordre non entier. Hermes, New Castle (1991)
7. Petráš, I.: Fractional-Order Nonlinear Systems: Modeling. Analysis and Simulation. Springer, Beijing (2011)
8. Podlubny, I.: Fractional Differential Equations: An Introduction to Fractional Derivatives, Fractional Differential Equations, to Methods of Their Solution and some of their Applications. Academic, San Diego (1999)
9. Podlubny, I.: Fractional-order systems and $PI^\lambda D^\mu$-controllers. IEEE Trans. Autom. Control **44**(1), 208–214 (1999)
10. Sabatier, J., Oustaloup, A., García-Iturricha, A., Lanusse, P.: CRONE control: principles and extension to time-variant plants with asymptotically constant coefficients. Nonlinear Dyn. **29**(1), 363–385 (2002)

Chapter 6
Observer-Based Fault Diagnosis for Fractional-Order Nonlinear Systems

In recent years, fractional calculus has attracted the interest of the control community in relation to those physical phenomena which have been modelled as fractional order (FO) dynamics. Some FO dynamics in botanical electrical impedances are analysed employing Bode and polar diagrams, leading to electrical circuit models revealing a FO behaviour [1]. Recent works introduce FO models to emulate the ultracapacitor dynamic and lithium-ion batteries [2, 3]. More examples of such class of dynamics have been reported in [4] and references therein. All this fractional models allow to explore new lines of researches, in fact, it is sensible to extend integer order methodologies.

FD has been studied for many years and has been an important topic for system monitoring, fault-tolerant and active disturbance rejection control [5–8], even now remains as an issue of great importance. In the case of integer order (IO) systems recent works dealing with this topic can be found. Nowadays, the FD is addressed by different techniques, observer-based diagnosis is one of the main approaches, e.g., the diagnosis of a proton exchange membrane fuel cells and the sensor fault diagnosis of an industrial gas turbine are tackled by means of a Takagi–Sugeno interval observer and an adaptive sliding mode observer [9, 10]. Besides, there exist a wide range of strategies to deal with this issue, to mention some additional examples: A robust neural-network-based actuator fault estimation in nonlinear systems and a fault diagnosis approach based on the Group Method of Data Handling technique are presented in [6, 11]. On the other hand, for fractional non-linear systems (FNLS) there are very limited reports on fault detection and diagnosis, thus causing shortages of real-time applications and researches about fault diagnosis-based fault-tolerant control or even state estimation, e.g., two methods for fault detection are presented in references [12, 13]; however, they are not reconstructing faults of the system, even though the diagnosis is extremely important because of the information that

© The Author(s), under exclusive license to Springer Nature Switzerland AG 2021
R. Martínez-Guerra et al., *Fault-tolerant Control and Diagnosis for Integer and Fractional-order Systems*, Studies in Systems, Decision and Control 328,
https://doi.org/10.1007/978-3-030-62094-3_6

provides, as the magnitude of the faults, the time when the system is under effects of faults and their location (actuators, sensors or both).

This chapter contributes to the theory of FNLS, the observe-based fault diagnosis problem and the issue of unmeasurable state estimation, therefore, we are introducing important tools with the purpose of dealing with practical problems in fractional systems. The main contribution of this chapter is to present a novel technique for fault diagnosis and state estimation for FNLS. This methodology is motivated by the short number of fractional observers unlike the many options of traditional IO observers. Recently, some fractional observers address the state estimation problem for a class of FNLS under unknown and/or uncertainties dynamics [14–16], however, none estimates these unknown dynamics, a clear advantage of this approach. This new strategy is based on two new definitions; the fractional algebraic observability (FAO) and the fractional algebraic diagnosability (FAD), these properties are directly related with the output measurements, since there depend on measurable outputs, known inputs and their fractional derivatives.

The fault diagnosis and state estimation are illustrated through four numerical simulation. They are three different cases of systems where is possible to reconstruct faults, unmeasurable state variables or both. In particular, one of the simulation presents results for the glucose-insulin fractional system [16, 17] which is derived from real data [18–20].

6.1 Preliminary Concepts

This section introduces some general concepts in order to ease the understanding of the proposed scheme. There are several definitions of a fractional derivative of order α ($0 < \alpha \leq 1$) [21–23]; for this chapter, due to the meaning of the initial conditions (same as in IO systems), the Caputo fractional operator is applied.

6.1.1 Caputo Fractional Derivative

Definition 6.1 ([23]) The Caputo fractional derivative of order $\alpha \in \mathbb{R}^+$ of a function x is defined as:

$$x(t)^{(\alpha)} = {}_{t_0}\mathscr{D}_t^{(\alpha)} x(t) \tag{6.1}$$

$$= \frac{1}{\Gamma(n-\alpha)} \int_0^t \frac{d^n x(\tau)}{d\tau^n} (t-\tau)^{n-\alpha-1} d\tau$$

where $n-1 \leq \alpha < n$; $\frac{d^n x(\tau)}{d\tau^n}$ is the nth derivative of x in the usual sense, $n \in \mathbb{N}$ and Γ is the gamma function defined as

$$\Gamma(z) = \int_0^\infty e^{-t} t^{z-1} dt$$

that converges in the right half of the complex plane $\Re(z) > 0$.

Now, a sequential operator [24] is defined as follows

$$\mathscr{D}^{(r\alpha)} x(t) = \underbrace{{}_{t_0}\mathscr{D}_t^{(\alpha)} {}_{t_0}\mathscr{D}_t^{(\alpha)} \cdots {}_{t_0}\mathscr{D}_t^{(\alpha)} {}_{t_0}\mathscr{D}_t^{(\alpha)}}_{r-time} x(t) \tag{6.2}$$

i.e., it is the Caputo fractional derivative of order α applied $r \in \mathbb{N}$ times, with $\mathscr{D}_t^{(0)} x(t) = x(t)$, note that if r = 1 then $\mathscr{D}_t^{(\alpha)} x(t) = x^{(\alpha)}$.

6.1.2 Mittag-Leffler Function

The Mittag-Leffler function [25] with two parameters is defined as:

$$E_{\alpha,\beta}(z) = \sum_{i=0}^\infty \frac{z^i}{\Gamma(\alpha i + \beta)}, \quad \Re(\alpha) > 0, \quad \Re(\beta) > 0 \tag{6.3}$$

This function is used to solve fractional differential equations as the exponential function in IO systems. Now if we have particular values of α, the function (6.3) has asymptotic behaviour at infinity.

Theorem 6.1 ([23]) *If $\alpha \in (0, 2)$, β is an arbitrary complex number and δ is an arbitrary real number such that*

$$\frac{\pi \alpha}{2} < \delta < \min(\pi, \pi\alpha), \tag{6.4}$$

then for an arbitrary integer $k \geq 1$ the following expansion holds:

$$E_{\alpha,\beta}(z) = -\sum_{i=0}^k \frac{1}{\Gamma(\beta - \alpha i) z^i} + O\left(\frac{1}{|z|^{k+1}}\right) \tag{6.5}$$

with $|z| \to \infty$, $\delta \leq |arg(z)| \leq \pi$.

Finally, two important properties of the Mittag-Leffler function are presented.

Property 6.1 ([23])

$$\int_0^t \tau^{\beta-1} E_{\alpha,\beta}(-k\tau^\alpha) d\tau = t^\beta E_{\alpha,\beta+1}(-kt^\alpha), \quad \beta > 0.$$

Property 6.2 ([24]) $E_{\alpha,\beta}(-x)$, *is completely monotonic, i.e.,* $(-1^n)E_{\alpha,\beta}^{(n)}(-x) \geq 0$ *for* $0 < \alpha \leq 1$ *and* $\beta \geq \alpha$ *for all* $x \in (0, \infty)$ *and* $n \in \mathbb{N} \cup \{0\}$.

6.2   Statement of the Diagnosis Problem

First, let us introduce the problem of FD. A FNLS with faults is described by the following equations:

$$x^{(\alpha)} = F(x, \bar{u}) \qquad 0 < \alpha \leq 1 \qquad (6.6)$$
$$y = h(x)$$

with $x^T = (x_1, x_2, \ldots, x_n) \in \mathbb{R}^n$ is the state vector, $\bar{u} = (u, f) = (u_1, u_2, \ldots, u_m, f_1, f_2, \ldots, f_\mu) \in \mathbb{R}^{m-\mu} \times \mathbb{R}^\mu$ where u is a known input vector and f is an unknown fault vector, $y = (y_1, y_2, \ldots, y_p) \in \mathbb{R}^p$ is the measured output.

Now the problem is: how can we estimate the unknown fault vector?, this question arises because, if we could measure the unknown input f, the problem of FD would be already solved and the information that provides could be used for decision-making (maintenance or replacement of actuators, sensors or both) and even for control issues. In order to solve this necessity, we can see it as an observation problem and introduce the following definitions.

Definition 6.2 (*FAO*) A state variable $x_i \in \mathbb{R}$ satisfies the FAO property, if it is a function of the firsts $r_1, r_2 \in \mathbb{N}$ sequential fractional derivatives (in the sense of the Eq. (6.2)) of the available output y and the known input vector u.

$$x_i = \phi_{x_i}(y, y^{(\alpha)}, \ldots, y^{(r_1\alpha)}, u, u^{(\alpha)}, \ldots, u^{(r_2\alpha)}). \qquad (6.7)$$

Definition 6.3 (*FAD*) A FNLS described by Eq. (6.6) is a fractional diagnosable system, if f_i meets Definition 6.2 with respect to y, u and the $r_1, r_2 \in \mathbb{N}$ sequential derivatives, i.e, f_i can be written as

$$f_i = \phi_{f_i}(y, y^{(\alpha)}, \ldots, y^{(r_1\alpha)}, u, u^{(\alpha)}, \ldots, u^{(r_2\alpha)}). \qquad (6.8)$$

To illustrate Definitions 6.2 and 6.3 the following examples are introduced.

Example 6.1 The FNLS described as

$$x_1^{(\alpha)} = x_1 x_2 + u + f$$
$$x_2^{(\alpha)} = x_1 \qquad 0 < \alpha \leq 1 \qquad (6.9)$$
$$y = x_2$$

is fractional diagnosable and fractional observable since x_1 and f satisfy the fractional polynomial equations given by

$$x_1 = y^{(\alpha)} \tag{6.10}$$
$$f = y^{(2\alpha)} - y^{(\alpha)}y - u \tag{6.11}$$

thus x_1 and f meet the FAO and FAD definitions respectively, allowing the estimation of the unmeasurable state x_1 and the unknown input f.

Remark 6.1 A fractional diagnosable system does not necessarily meet Definition 6.2, and vice versa. Indeed, the following example illustrates this fact, when a system is fractional diagnosable, but not a fractional observable system.

Example 6.2 Let us consider the following FNLS

$$\begin{aligned} x_1^{(\alpha)} &= x_2 \\ x_2^{(\alpha)} &= -x_2 + f \qquad\qquad 0 < \alpha \le 1 \\ x_3^{(\alpha)} &= x_1 x_3 - x_2 \\ y &= x_2. \end{aligned} \tag{6.12}$$

It is not hard to see that (6.12) is a fractional diagnosable system according to Definition 6.3 giving the possibility of estimate f from $y^{(\alpha)}$ and y,

$$f = y^{(\alpha)} + y$$

however, x_1 and x_3 are not fractional algebraically observable precluding the state estimation.

It is clear that Definitions 6.2 and 6.3 are playing an important role in the proposed methodology, since they allow to know beforehand whether faults, unmeasurable state variables or both can be estimated.

6.3 Fractional Reduced-Order Observer

In this section an upper bound for the solution of the fractional error dynamic and a special structure of the proposed fractional observer. Firstly, let us consider the system (6.6), the unknown fault vector f can be interpreted as a state with uncertain fractional dynamic. In order to estimate f, the state vector is extended to deal with the unknown vector. The new extended system is given by

$$\begin{aligned} x^{(\alpha)} &= F(x, \bar{u}) \\ f^{(\alpha)} &= \Omega(x, \bar{u}) \qquad\qquad 0 < \alpha \le 1 \\ y &= h(x). \end{aligned} \tag{6.13}$$

where $\Omega(\bullet)$ is a bounded uncertain function.

Then, the problem is overcome by using a fractional reduced-order observer in order to reconstruct the fault vector f and possibly estimate state variables.

Lemma 6.1 *If the following hypotheses are satisfied:*

H1 : $\Omega_i(x, \bar{u})$ *is bounded, i.e.* $\|\Omega_i(x, \bar{u})\| \leq M_i$, $M_i \in \mathbb{R}^+$, $0 < i \leq \mu$
H2 : $f_i(t)$ *satisfies the FAD condition*
H3 : $K_{\hat{f}_i} \in \mathbb{R}^+$

then, the system

$$\hat{f}_i^{(\alpha)} = K_{\hat{f}_i}(f_i - \hat{f}_i) \quad 0 < i \leq \mu, \quad 0 < \alpha \leq 1 \tag{6.14}$$

is a fractional reduced-order observer which achieves the diagnosis of the fault f_i, where $\hat{f}_i$ denotes the estimated fault, and $K_{\hat{f}_i}$ determines a desired convergence rate for the fractional observer.

Proof The fractional dynamic for $e_{f_i} = f_i - \hat{f}_i$ can be expressed as

$$e_{f_i}^{(\alpha)} + K_{\hat{f}_i} e_{f_i} = \Omega_i(x, \bar{u}) \tag{6.15}$$

There exist a unique solution [25] for the system (6.15), due to $\Omega_i(x(t), \bar{u}(t)) - K_{\hat{f}_i} e(t)_{\hat{f}_i}$ is a Lipschitz continuous function on e. Solving Eq. (6.15), the following equation is given

$$e_{f_i}(t) = e_{f_{i0}} E_{\alpha,1}(-K_{\hat{f}_i} t^\alpha)$$
$$+ \int_0^t (t - \tau)^{\alpha-1} E_{\alpha,\alpha}(-K_{\hat{f}_i}(t - \tau)^\alpha) \Omega_i(x(\tau), \bar{u}(\tau)) d\tau$$

where $e_{f_i}(0) = e_{f_{i0}}$.

Using triangle and Cauchy–Schwarz inequalities and **H1** we have

$$|e_{fi}(t)| \leq |e_{f_{i0}} E_{\alpha,1}(-K_{\hat{f}_i} t^\alpha)|$$
$$+ M_i \int_0^t |(t - \tau)^{\alpha-1} E_{\alpha,\alpha}(-K_{\hat{f}_i}(t - \tau)^\alpha)| d\tau \tag{6.16}$$

The functions $E_{\alpha,1}(-K_{\hat{f}_i} t^\alpha)$ and $(t - \tau)^{\alpha-1} E_{\alpha,\alpha}(-K_{\hat{f}_i}(t - \tau)^\alpha)$ are not negative due to **Property** 6.2 of Mittag-Leffler function and **H3**. Then

$$|e_{fi}(t)| \leq |e_{f_{i0}}| E_{\alpha,1}(-K_{\hat{f}_i} t^\alpha)$$
$$+ M_i \int_0^t (t - \tau)^{\alpha-1} E_{\alpha,\alpha}(-K_{\hat{f}_i}(t - \tau)^\alpha) d\tau \tag{6.17}$$

Applying **Property** 6.1 of Mittag-Leffler function

$$|e_{fi}(t)| \leq |e_{fi0}|E_{\alpha,1}(-K_{\hat{f}_i}t^\alpha) + M_i t^\alpha E_{\alpha,\alpha+1}(-K_{\hat{f}_i}t^\alpha) \tag{6.18}$$

If $t \to \infty$, we use Eq. (6.5) with $\delta = \frac{3\pi\alpha}{4}$ due to **H3**

$$\lim_{t\to\infty}|e_{f_i}| \leq |e_{fi0}|\lim_{t\to\infty}E_{\alpha,1}(-K_{\hat{f}_i}t^\alpha) + M_i \lim_{t\to\infty}t^\alpha E_{\alpha,\alpha+1}(-K_{\hat{f}_i}t^\alpha)$$
$$\leq R \tag{6.19}$$

hence $e_{f_i} \in \mathscr{B}_R(0)$, with $R = \frac{M_i}{K_{\hat{f}_i}}$. $\qquad\qquad\qquad\qquad\qquad\qquad\square$

Remark 6.2 Sometimes the fractional derivatives (unknown signals) of measurable outputs appear in the polynomial fractional algebraic equation (FAD condition), hence, it is necessary to use auxiliary variables to avoid the estimation of them.

Corollary 6.1 *If a fault signal f_i with $1 \leq i \leq \mu$ meets the FAD condition, and the fault can be written as*

$$f_i = \sigma_i y_i^{(\alpha)} + \psi_i(y, u) \tag{6.20}$$

with $\sigma_i = (\sigma_{i1}, \ldots, \sigma_{ip}) \in \mathbb{R}^p$ being a constant vector, y and $\psi_i(y, u)$ bounded functions, then, the fractional reduced-order observer (6.14) can be rewritten as follows:

$$\gamma_{f_i}^{(\alpha)} = -K_{\hat{f}_i}(\gamma_{f_i} - \psi_i(y, u)) - K_{\hat{f}_i}^2 \sigma_i y_i, \quad 0 < \alpha \leq 1 \tag{6.21}$$

$$\hat{f}_i = \gamma_{f_i} + K_{\hat{f}_i}\sigma_i y_i \tag{6.22}$$

where $\gamma_{f_i}(0) = \gamma_{f_{i0}}$.

Proof Considering Eqs. (6.20), (6.21) and (6.22), the fractional reduced-order observer (6.14) can be written as

$$\hat{f}_i^{(\alpha)} = K_{\hat{f}_i}(\underbrace{\sigma_i y_i^{(\alpha)} + \psi_i(y, u)}_{f_i} - \hat{f}_i) \tag{6.23}$$

Now, let us define γ_{f_i} as follows

$$\gamma_{f_i} \triangleq \hat{f}_i - K_{\hat{f}_i}\sigma_i y_i \tag{6.24}$$

and taking the fractional derivative of (6.24), we get

$$\gamma_{f_i}^{(\alpha)} = \hat{f}_i^{(\alpha)} - K_{\hat{f}_i}\sigma_i y_i^{(\alpha)} \tag{6.25}$$

Finally, substituting (6.23) in (6.25) yields

$$\gamma_{f_i}^{(\alpha)} = -K_{\hat{f}_i}(\gamma_{f_i} - \psi_i(y, u)) - K_{\hat{f}_i}^2 \sigma_i y_i, \;\; \gamma_{f_i}(0) = \gamma_{f_{i0}} \tag{6.26}$$

thus, Eq. (6.26) is obtained and the proof is completed. $\square$

Remark 6.3 Lemma 6.1 and Corollary 6.1 are equivalent for the case of estimating state variables due to the realization between the definitions; FAO and FAD.

The performance of the fractional reduced-order observer is analysed by means of four numerical simulations. The four FNLS have special characteristics in order to illustrate the versatility of the proposed methodology. The first example requires a simultaneous estimation of faults, i.e, the system is under effect of multiple faults. The following two systems satisfy the FAO and FAD conditions, which allows to reconstruct faults and states considering linear and nonlinear outputs. Finally, the dynamic response of a diabetic patient's blood glucose concentration to the insulin injection, that is under effects of an unknown periodic function, is successfully diagnosed.

6.4   Numerical Simulations

In order to verify the effectiveness of the proposed method, some numerical simulations were carried out. All the simulations were performed using Matlab® Simulink Ninteger Toolbox [26] and implemented with a sample time of 0.0001 s. First, consider the following FNLS, which requires a simultaneous FD

$$\begin{aligned}
x_1^{(\alpha)} &= -x_1 + f_1 x_2^3 + f_2 x_2 x_3 + u(t) \\
x_2^{(\alpha)} &= x_3 + f_1 \\
x_3^{(\alpha)} &= -x_2^3 + f_2 \\
y_1 &= x_2 \\
y_2 &= x_3
\end{aligned} \tag{6.27}$$

where $x = (x_1, x_2, x_3)^T$ is the state vector, $f = (f_1, f_2)^T$ is the fault vector, $y = (y_1, y_2)^T$ is the output vector and $u(t)$ is a known input.

Based on the proposed methodology, the FAD condition for f_1 and f_2 can be obtained by considering the dynamic equations of x_2 and x_3 from system (6.27), i.e, the unknown signals (or faults) are written in terms of measurable outputs and their fractional derivatives.

$$f_1 = \phi_{f_1}(y, y^{(\alpha)}) = y_1^{(\alpha)} - y_2 \tag{6.28}$$

$$f_2 = \phi_{f_2}(y, y^{(\alpha)}) = y_2^{(\alpha)} + y_1^3 \tag{6.29}$$

Then, from Eq. (6.14), and considering (6.28) and (6.29), the fractional reduced-order observers for f_1 and f_2 are described as follows:

$$\hat{f}_1^{(\alpha)} = K_{\hat{f}_1} (\underbrace{y_1^{(\alpha)} - y_2}_{f_1} - \hat{f}_1)$$

$$\hat{f}_2^{(\alpha)} = K_{\hat{f}_2} (\underbrace{y_2^{(\alpha)} + y_1^3}_{f_2} - \hat{f}_2)$$

Note that the signals $y_1^{(\alpha)}$ and $y_2^{(\alpha)}$ are not available, i.e, they are not measurable outputs. However, taking into account the methodology presented in Corollary 6.1, the following two artificial variables can be defined in order to avoid the need to estimate any fractional derivative.

$$\gamma_{f_1} \triangleq \hat{f}_1 - K_{\hat{f}_1} y_1 \tag{6.30}$$

$$\gamma_{f_2} \triangleq \hat{f}_2 - K_{\hat{f}_2} y_2 \tag{6.31}$$

Finally, taking the fractional derivative of (6.30) and (6.31), the fractional reduced-order observers are rewritten as:

$$\gamma_{f_1}^{(\alpha)} = -K_{\hat{f}_1}(\gamma_{f_1} + y_2) - K_{\hat{f}_1}^2 y_1, \quad \gamma_{f_1}(0) = \gamma_{f_{10}} \tag{6.32}$$

$$\hat{f}_1 = \gamma_{f_1} + K_{\hat{f}_1} y_1$$

$$\gamma_{f_2}^{(\alpha)} = -K_{\hat{f}_2}(\gamma_{f_2} - y_1^3) - K_{\hat{f}_2}^2 y_2, \quad \gamma_{f_2}(0) = \gamma_{f_{20}} \tag{6.33}$$

$$\hat{f}_2 = \gamma_{f_2} + K_{\hat{f}_2} y_2.$$

Remark 6.4 Note that the fractional observers (6.32) and (6.33) are now independent of unknown signals. Due to the methodology introduced by Corollary 6.1, it was possible to rewrite the fractional observers in terms of artificial variables and measurable outputs. As result, the need to estimate unknown fractional derivatives was avoided.

The numerical simulation for the system (6.27) is carried out considering the dynamics of the faults f_1 and f_2, described by the following equations:

$$f_1 = 1 + 5\sin(x_1)e^{-10t}$$

$$f_2 = 5e^{-5(t-0.3)}\mathscr{U}(t - 0.3) + e^{-5(t-1)}\mathscr{U}(t - 1)$$

where $\mathscr{U}(t)$ is the unit step function. The simulation results are obtained by considering the initial conditions $\gamma_{f_{10}} = \gamma_{f_{20}} = 0$ and the parameters $K_{\hat{f}_1}$, $K_{\hat{f}_2}$, α and the known input $u(t)$ as $K_{\hat{f}_1} = K_{\hat{f}_2} = 40$, $\alpha = 0.935$ and $u(t) = 5\mathscr{U}(t)$. The results are displayed in Figs. 6.1, 6.2, 6.3 and 6.4. It can be noted that in Figs. 6.1 and 6.2, the

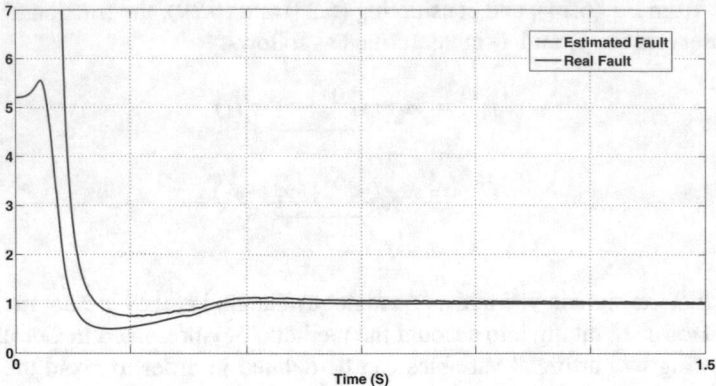

Fig. 6.1 Fault reconstruction without output noise measurement for f_1 using a fractional reduced-order observer

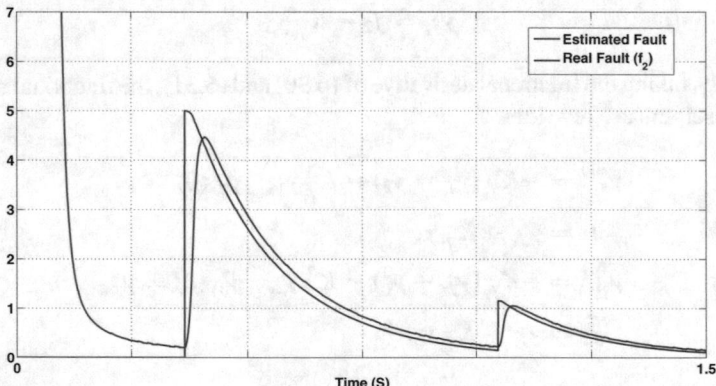

Fig. 6.2 Fault reconstruction without output noise measurement for f_2 using a fractional reduced-order observer

estimated faults follow their corresponding true value, two types of fault are used in order to show the versatility of the fractional observer. Clearly, the estimation of smooth functions f_1 shows better estimation compared to non differentiable functions f_2. However, with a precise tune of the parameters of the observer, it is possible to reconstruct this type of faults.

Figures 6.3 and 6.4 show the estimation of the two faults even in presence of measurement noise. The output measurements are contaminated with Gaussian noise, with zero mean, variance 0.01 and is bounded within the interval $[-0.001, 0.001]$.

Remark 6.5 For real-time applications, the use of a filter to reduce measurement output noise is recommended.

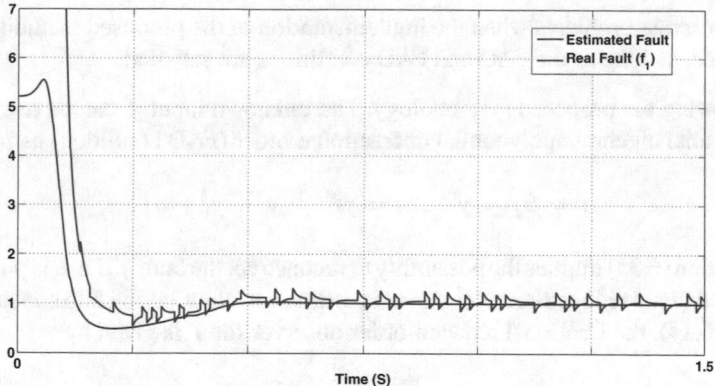

Fig. 6.3 Fault reconstruction for f_1 in presence of output noise measurement using a fractional reduced-order observer

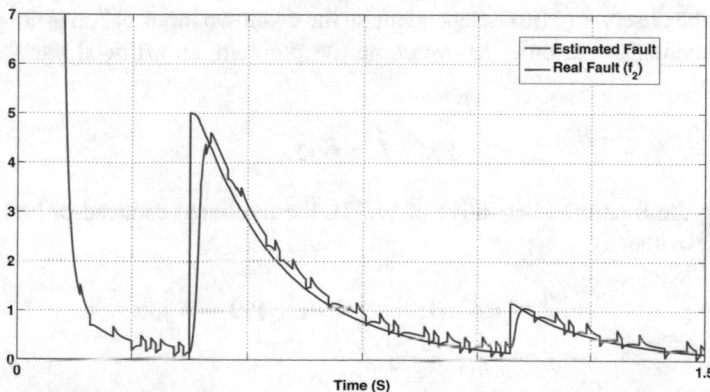

Fig. 6.4 Fault reconstruction for f_2 in presence of output noise measurement using a fractional reduced-order observer

Now, let us consider the FNLS described by the following equations:

$$
\begin{aligned}
x_1^{(\alpha)} &= x_1 x_2 + x_1^2 + f + u_1 \\
x_2^{(\alpha)} &= x_1 + u_2 \\
y_1 &= x_1 x_2 \\
y_2 &= x_1
\end{aligned}
\tag{6.34}
$$

where $x = (x_1, x_2)^T$ is the state vector, f is the unknown input (or fault), $y = (y_1, y_2)^T$ is the output vector and $u = (u_1, u_2)^T$ is the known input vector.

Remark 6.6 The system (6.34) shows a particularity to be highlighted. It is considered the case in which there is a nonlinear output. This type of measurements do not generate problems when the implementation of the proposed methodology is performed, as long as the FAO and FAD conditions are satisfied.

Following the proposed methodology. The unknown input f can be rewritten as a differential algebraic polynomial of fractional order (FAD condition) as follows.

$$f = \phi_f(y, y^{(\alpha)}, u) = y_2^{(\alpha)} - y_1 - y_1^2 - u_1 \tag{6.35}$$

Equation (6.35) implies the possibility to reconstruct the fault f, i.e, it is possible to built a fractional reduced-order observer to estimate it. Now, taking into consideration the Eq. (6.14), the fractional reduced-order observer for f is given by

$$\hat{f}^{(\alpha)} = K_{\hat{f}}(\underbrace{y_2^{(\alpha)} - y_1 - y_1^2 - u_1}_{f} - \hat{f}) \tag{6.36}$$

Since the observer (6.36) is dependent on the unknown input $y_2^{(\alpha)}$, it is not possible its implementation. In order to overcome the problem, an artificial variable γ_f is introduced.

$$\gamma_f \triangleq \hat{f} - K_{\hat{f}} y_2 \tag{6.37}$$

Taking the fractional derivative of (6.37), the fractional reduced-order observer for f is rewritten as

$$\gamma_f^{(\alpha)} = -K_{\hat{f}}(y_1 + y_1^2 + u_1 + \gamma_f) - K_{\hat{f}}^2 y_2 \tag{6.38}$$
$$\hat{f} = \gamma_f + K_{\hat{f}} y_2$$

where $\gamma_f(0) = \gamma_{f_0}$. Equation (6.38) shows a fractional observer which does not depend on $y_2^{(\alpha)}$, this observer is easy to implement since it is not necessary to estimate any unknown signal.

To obtain the simulation results, it is considered a fault f with the following dynamic:

$$f = (4\sin(0.5t) + 2\cos(t))\mathcal{U}(t-4) \tag{6.39}$$

f is a periodic smooth function and according to Eq. (6.39) the effects of the fault begin in the fourth second of the process. Until the effects of f do not appear the observer will reach zero indicating the absence of any malfunction or degradation of the system. In Fig. 6.5 it is shown that after four seconds the system (6.34) is affected by the fault, then the fractional observer starts the estimation process.

The parameters selected for the observer implementation are: $\gamma_{f_0} = 0$ as initial condition, the constant $K_{\hat{f}} = 15$ and the order of the fractional derivative $\alpha = 0.935$.

The two previous examples showed the versatility of the proposed methodology for the estimation of faults. The example below illustrates the implementation of

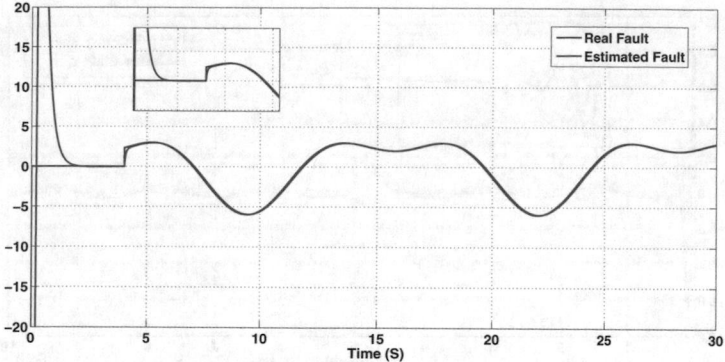

Fig. 6.5 Fault reconstruction for f using a fractional reduced-order observer

the methodology in order to estimate unknown states and faults. In Sect. 6.2 was mentioned that the system (6.9) meets the FAO and FAD conditions, i.e, it is possible to design fractional reduced-order observers for the estimation of the unmeasurable state x_1 and the fault f.

Equation (6.10) is considered for the construction of the fractional observer for x_1 which is described as

$$\gamma_{x_1}^{(\alpha)} = -K_{\hat{x}_1}\gamma_{x_1} - K_{\hat{x}_1}^2 y \qquad \gamma_{x_1}(0) = \gamma_{x_{10}}$$
$$\hat{x}_1 = \gamma_{x_1} + K_{\hat{x}_1}$$

For the observer of the fault the methodology presented in Corollary 6.1 becomes iterative. Since (6.11) is dependent on $y^{(2\alpha)}$ a new artificial variable $\eta \triangleq y^{(\alpha)}$ must be defined in order to avoid to reconstruct $y^{(2\alpha)}$. Clearly, the number of artificial variables is linked to the maximum order of the factional polynomial.

$$\gamma_{\eta}^{(\alpha)} = -K_{\hat{\eta}}\gamma_{\eta} - K_{\hat{\eta}}^2 y, \quad \gamma_{\eta}(0) = \gamma_{\eta_0}$$
$$\gamma_f^{(\alpha)} = -K_{\hat{f}}((\gamma_{\eta} + K_{\hat{\eta}}y)(y + K_{\hat{f}}) + u + \gamma_f), \quad \gamma_f(0) = \gamma_{f_0}$$
$$\hat{f} = \gamma_f + K_{\hat{f}}(\gamma_{\eta} + K_{\hat{\eta}}y)$$

The gains $K_{\hat{x}_1}$, $K_{\hat{\eta}}$, $K_{\hat{f}}$ and the parameter α are chosen as $K_{\hat{x}_1} = 20$, $K_{\hat{\eta}} = 25$, $K_{\hat{f}} = 30$ and $\alpha = 0.935$. The simulation results with initial conditions as $\gamma_{x_{10}} = \gamma_{\eta_0} = \gamma_{f_0} = 0$ are presented in Figs. 6.5 and 6.6, they display the state estimation and a successful fault diagnosis. For this simulation the real dynamic of the fault is given by

$$f = 3\sin(4t)$$

Furthermore, Bergman's generalized minimal model for glucose-insulin is a commonly referenced model that approximates the dynamic response of a diabetic

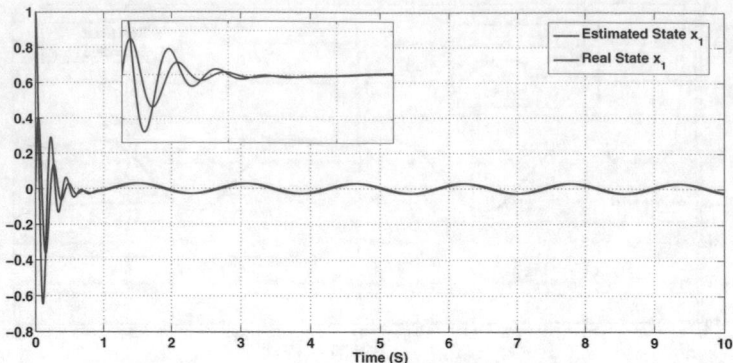

Fig. 6.6 State reconstruction for x_1 using a fractional reduced-order observer

patient's blood glucose concentration to the insulin injection [18–20]. For this section, let us consider the FO model that monitors the temporal dynamics of blood glucose concentration [16],

$$
\begin{aligned}
x_1^{(\alpha)} &= -p_1(x_1 - G_b) - x_1 x_2 + f(t) \\
x_2^{(\alpha)} &= -p_2 x_2 + p_3(x_3 - I_b) \qquad 0 < \alpha < 1 \\
x_3^{(\alpha)} &= -n(x_3 - I_b) + u \\
y &= x_1
\end{aligned}
\tag{6.40}
$$

where x_1, x_2 and x_3 are the blood plasma glucose concentration (mg/dl), the insulin effect on the net glucose disappearance (1/min) and the insulin concentration in plasma (μU/ml) respectively, the glucose concentration in the blood is considered as the output y. G_b represents the basal pre-injection level of glucose, I_b is the basal pre-injection level of insulin and u acts as the control variable, replacing the normal insulin regulation of the body. The normal insulin regulation does not exist in diabetic patients, this glucose absorption is considered as a fault $f(t)$ for the system (6.40). The fault can be modeled by $A\sin(\omega t)$, representing circadian rhythms (endocrine cycles) with period of 8 h and amplitude around 10 mg/dl.

$$
f(t) = A\sin(\omega t)
$$

with $A = 10$ mg/dl, $\omega = \frac{2\pi}{T} \frac{\text{rad}}{\text{s}}$ and $T = 8$ h.

For this example the fault component is given by

$$
f = \phi_f(y, y^{(\alpha)}, x_2) = y^{(\alpha)} + p_1(y - G_b) + y x_2
\tag{6.41}
$$

where G_b and p_1 are known constants. It is not hard to see that f does not meet the definition of FAD due to dependence on x_2, to solve this problem we can define the state variables $x_2 = \beta$ and $x_3 = \xi$ and rewrite the system (6.40) as

$$x_1^{(\alpha)} = -p_1(x - G_b) - \beta x_1 + f(t), \quad 0 < \alpha < 1 \tag{6.42}$$
$$y = x_1$$

where, β and ξ have dynamics described by

$$\beta^{(\alpha)} + p_2\beta = \Psi(\xi) \qquad\qquad 0 < \alpha < 1 \tag{6.43}$$
$$\xi^{(\alpha)} + n\xi = \Lambda(u) \tag{6.44}$$

with $\Psi = -p_3(\xi - I_b)$ and $\Lambda = nI_b + u$.

Solving the Eqs. (6.43) and (6.44) [25], the following equations are given

$$\xi(t) = \xi_0 E_{\alpha,1}(-nt^\alpha)$$
$$+ \int_0^t (t - \tau)^{\alpha-1} E_{\alpha,\alpha}(-n(t - \tau)^\alpha)\Lambda(u(\tau))d\tau$$
$$\beta(t) = \beta_0 E_{\alpha,1}(-p_2 t^\alpha)$$
$$+ \int_0^t (t - \tau)^{\alpha-1} E_{\alpha,\alpha}(-p_2(t - \tau)^\alpha)\Psi(\xi(\tau))d\tau$$

where $\xi(0) = \xi_0$, $\beta(0) = \beta_0$.

Finally, with (6.41) and the solutions of ξ and β, the following fractional equation is obtained.

$$f^{(\alpha)} = K_{\hat{f}}(\underbrace{y^{(\alpha)} + p_1(y - G_b) + y\beta(t) - \hat{f}}_{f})$$

Using Corollary 6.1 and selecting the variable $\gamma_f \triangleq \hat{f} - K_{\hat{f}}y$ the fractional reduced-order observer for f in the system (6.40) is defined as

$$\gamma_f^{(\alpha)} = -K_{\hat{f}}(\gamma_f - p_1(y - G_b) - y\beta(t)) - K_{\hat{f}}^2 y$$
$$\hat{f} = \gamma_f + K_{\hat{f}}y$$

The numerical results are obtained with known constants $p_1 = 0$, $p_2 = 0.02$, $p_3 = 5.3 \times 10^{-6}$, $n = 0.3$, $G_b = 70$, $I_b = 7$ and initial conditions $x_{10} = 220$, $x_{20} = 0$ and $x_{30} = 50$. The fractional observer gain $K_{\hat{f}}$ and the initial condition are taken as $K_{\hat{f}} = 10$, $\gamma_{f0} = 0$. The simulation was performed for 800 s and Figs. 6.7, 6.8, 6.9, 6.10 and 6.11 show how the diagnosis of the fault is achieved, particularly, Fig. 6.9 illustrates the first 30 s where the fault has a high frequency that hinders the diagnosis, however, a few seconds later, the observer achieves the diagnosis of the fault.

Another interesting result is illustrated in Fig. 6.11, this figure shows when the fault is not acting since the begin of simulation, the fault appears after 100 s meanwhile the algorithm tries to reach zero, finally, when the system is under effects of faults, the observer begins the diagnostic or monitoring process.

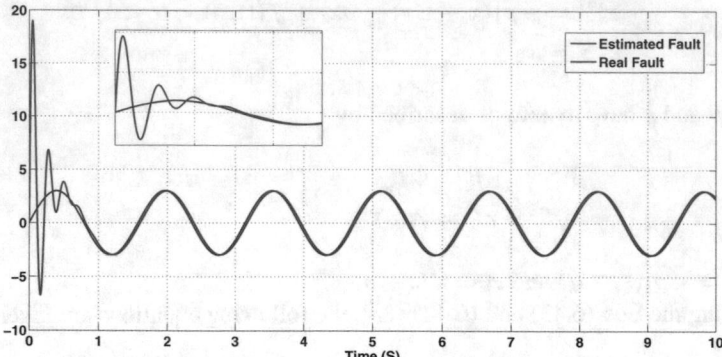

Fig. 6.7 Fault reconstruction for f using a fractional reduced-order observer

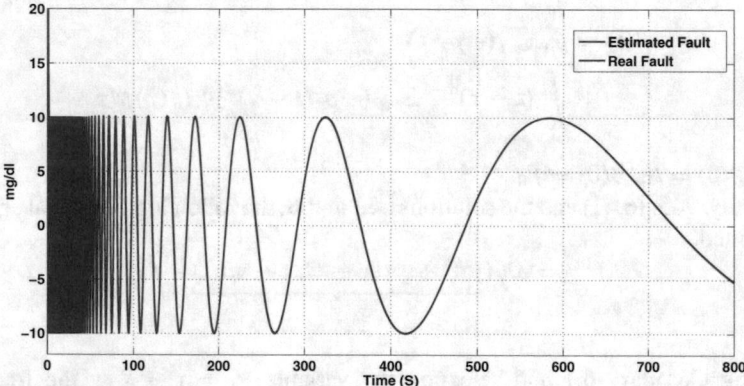

Fig. 6.8 Fault reconstruction for f in a period of 800 s

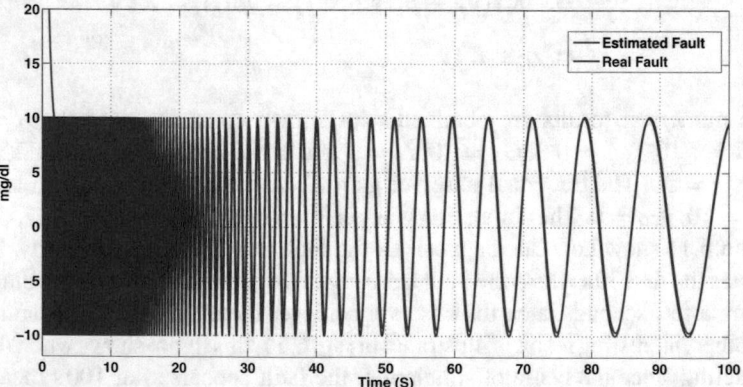

Fig. 6.9 Fault reconstruction for f in a period of 100 s

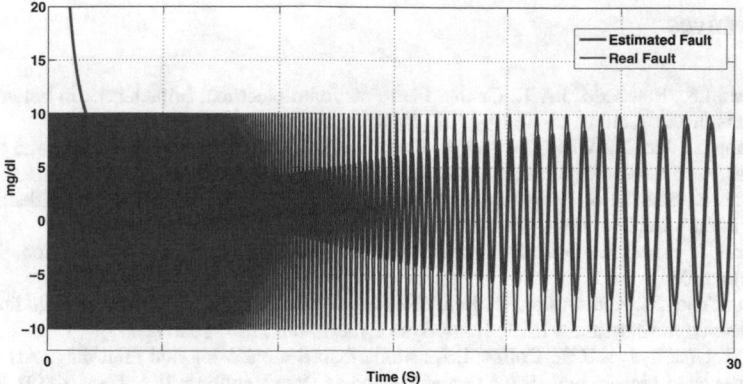

Fig. 6.10 Fault reconstruction for f in a period of 30 s

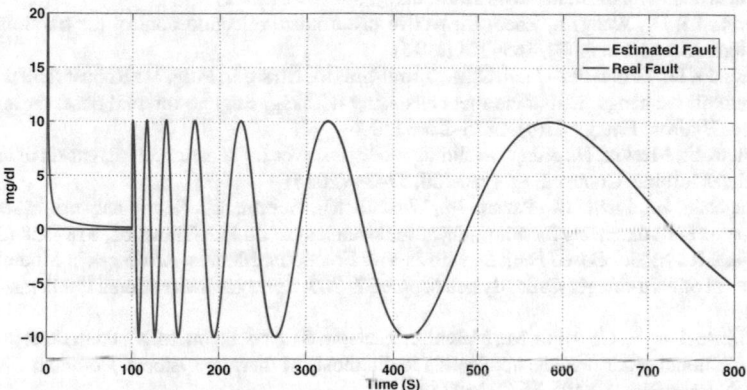

Fig. 6.11 Fault reconstruction for f in presence of the fault in a period of 100–800 s

6.5 Concluding Remarks

A novel fault diagnosis scheme applied to FNLS was addressed by means of a fractional reduced-order observer. This diagnostic methodology was capable of estimating multiple faults and unmeasurable state variables. The definitions of fractional algebraic observability and the fractional algebraic diagnosability were introduced with the purpose of ensuring the estimation of the unknown signals (state variables and faults). Working under these assumptions, four simulations were performed to reveal the advantages of the presented methodology. Finally, The Bergman generalized minimal mode glucose-insulin model was presented and successfully diagnosed in order to motivate real-time applications using fractional techniques.

References

1. Jesus, I.S., Machado, J.A.T., Cunha, J.B.: Fractional electrical impedances in botanical elements. J. Vib. Control **14**, 1389–1402 (2008)
2. Zhang, L., Hu, X., Wang, Z., Sun, F., Dorrell, D.G.: Fractional-order modeling and State-of-Charge estimation for ultracapacitors. J. Power Sources **314**, 28–34 (2016)
3. Wang, B., Eben Li, S., Peng, H., Liu, Z.: Fractional-order modeling and parameter identification for lithium-ion batteries. J. Power Sources **293**, 151–161 (2015)
4. Petras, I.: Fractional-Order Nonlinear Systems: Modeling, Analysis and Simulation. Springer, Berlin (2011)
5. Feng Zhao, X., Koutsoukos, H., Reich Haussecker, J., Cheung, P.: Monitoring and fault diagnosis of hybrid systems. IEEE Trans. Syst. Cybern. **35**, 1225–1240 (2005)
6. Li, F., Upadhyaya, B.R., Coffey, L.A.: Model-based monitoring and fault diagnosis of fossil power plant process units using Group Method of Data Handling. ISA Trans. **48**(2), 213–219 (2009)
7. Wang, T., Xie, W., Zhang, Y.: Sliding mode fault tolerant control dealing with modeling uncertainties and actuator faults. ISA Trans. **51**(3), 386–392 (2012)
8. Li, M., Li, D., Wang, J., Zhao, C.: Active disturbance rejection control for fractional-order system. ISA Trans. **52**(3), 365–374 (2013)
9. Rotondo, D., Fernandez-Canti, R.M., Tornil-Sin, S., Blesa, J., Puig, V.: Robust fault diagnosis of proton exchange membrane fuel cells using a Takagi-Sugeno interval observer approach. Int. J. Hydrog. Energy **41**(4), 2875–2886 (2016)
10. Rahme, S., Meskin, N.: Adaptive sliding mode observer for sensor fault diagnosis of an industrial gas turbine. Control Eng. Pract. **38**, 57–74 (2015)
11. Mrugalski, M., Luzar, M., Pazera, M., Witczak, M., Aubrun, C.: Neural network-based robust actuator fault diagnosis for a non-linear multi-tank system. ISA Trans. **61**, 318–328 (2016)
12. Kopka, R.: Model-Based Fault Diagnosis with Fractional Models, Advances in Modelling and Control of Non-integer-Order Systems. pp. 257–263. Springer International Publishing, Berlin (2015)
13. Aribi, A., Farges, C., Aoun, M., Melchior, P., Najar, S., Abdelkrim, M.N.: Fault detection based on fractional order models: application to diagnosis of thermal systems. Commun. Nonlinear Sci. Numer. Sim. **19**(10), 3679–3693 (2014)
14. Tamhane, B., Mujumdar, A., Kurode, S.: Fractional order disturbance observer based robust control. In: 2015 International Conference on Industrial Instrumentation and Control (ICIC), Pune, pp. 1412–1417 (2015)
15. Zhong, F., Li, H., Zhong, S.: State estimation based on fractional order sliding mode observer method for a class of uncertain fractional-order nonlinear systems. Signal Process. **127**, 168–184 (2016)
16. N'Doye, I., Voos, H., Darouach, M., Schneider, J.G., Knauf, N.: An unknown input fractional-order observer design for fractional-order glucose-insulin system. In: 2012 IEEE EMBS Conference on Biomedical Engineering and Sciences, Langkawi, pp. 595–600 (2012)
17. N'Doye, I., Voos, H., Darouach, M., Schneider, J.G.: Static output feedback $\mathscr{H}_\infty$ control for factional-order glucose-insuline system. Int. J. Control Autom. Syst. **13**(4), 798–807 (2015)
18. Neatpisarnvanit, C., Boston, J.: Estimation of plasma insulin from plasma glucose. IEEE Trans. Biomed. Eng. **49**, 1253–1259 (2002)
19. Faruque Ali, S., Padhi, R.: Optimal blood glucose regulation of diabetic patients using single network adaptive critics. Opt. Control Appl. Methods **32**, 196–214 (2009)
20. Markakis, M.G., Mitsis, G.D., Papavassilopoulos, G.P., Ioannou, P.A., Marmarelis, V.Z.: A switching control strategy for attenuation of blood glucose disturbances. Opt. Control Appl. Methods **32**, 185–195 (2011)
21. Podlubny, I.: Geometric and physical interpretation of fractional integration and fractional differentiation. Fract. Calculus Appl. Anal. **5**(4), 367–386 (2002)
22. Oldham, K., Spanier, J.: The Fractional Calculus: Theory and Applications of Differentiation and Integration of Arbitrary Order. Dover Books on Mathematics (2006)

23. Podlubny, I.: Fractional Differential Equations. Academic, New York (1999)
24. Miller, K.S., Ross, B.: An Introduction to the Fractional Calculus and Fractional Differential Equations. Wiley, New York (1993)
25. Kilbas, A.A., Srivastava, H.M., Trujillo, J.J.: Theory and Applications of Fractional Differential Equations. Elsevier B.V, Amsterdam (2006)
26. Valerio, D.: Ninteger v. 2.3 Fractional control toolbox for Matlab, User and Programmer Manual (2005)

Chapter 7
Fractional Integral Reduced-Order Observer

The main topic of this chapter is to present the fractional proportional integral reduced-order observer (FIROO). Similar to the PIROO, this observer will be used for fault diagnosis for fractional-order systems. The FIROO is shown to be Mittag-Leffler stable; its design is based on the fractional algebraic observability property.

To show the capabilities of the FIROO, it is used in this chapter to deal with the synchronization and anti-synchronization problems in commensurate and incommensurate-order fractional chaotic systems by means of the master-slave configuration. The observer (slave) estimates the states of the master system and synchronizes with it. It is also used to estimate some fractional derivatives of the output that appear in the dynamics. This methodology is applied in the fractional Lorenz chaotic system with commensurate dynamics, i.e. where the fractional order of the dynamics of all the states is the same, and in the fractional Rössler chaotic system with incommensurate dynamics. After applying the scheme, simulations are performed in order to obtain numerical results.

7.1 The Synchronization and Anti-synchronization Problems

Synchronization occurs when oscillatory (or repetitive) systems, via some kind of interaction, adjust their behaviours relative to one another so as to attain a state where they work in unison. Synchronization of oscillations is a phenomenon that was originally discovered by Christian Huygens in 1665 in a mechanical system: two pendulum "grandfather" clocks hanging on the same beam. Synchronization is encountered in various fields of science, in engineering and in social behaviour [4, 34].

© The Author(s), under exclusive license to Springer Nature Switzerland AG 2021   109
R. Martínez-Guerra et al., *Fault-tolerant Control and Diagnosis for Integer and Fractional-order Systems*, Studies in Systems, Decision and Control 328,
https://doi.org/10.1007/978-3-030-62094-3_7

Furthermore, chaotic systems have been of great interest in recent years due to their presence in physical systems of interest, such as mechanical, electrical and meteorological systems [19]. While describing the oscillation of chaotic systems in the phase space, it is found that it does not correspond to simple geometrical objects like a limit cycle, but rather to complex structures that are called strange attractors (in contrast to limit cycles that are simple attractors) [37]. The synchronization problem of chaotic systems has been extensively studied, given their applications to areas such as medicine [34], biological systems [3, 16] and secure communications hbox[23, 30, 32].

One of the methods used for solving the synchronization problem in chaotic systems consists in decomposing the system in question in two subsystems, where one of them acts as a driver in order to synchronize the behaviour of the other one. This approach requires the system to meet certain characteristics related to the Lyapunov coefficients, and thus it can be applied only to a certain kind of chaotic oscillators [36].

Another approach used is the so-called synchronization by observation, also known as master-slave configuration. In this case the chaotic system is taken as a "master" and an observer is designed, which estimates the dynamics of its unknown states by means of its output; this is why the observer is known as "slave" system. Different kinds of observers have been used as slave systems, such as the Luenberger [8, 38], high-gain [10], adaptive [14], reduced-order [45], polynomial [28] and bounded error [29]. Recently, the differential algebraic approach has been used for the design of reduced-order observers [31]. Given that these observers do not require to have full knowledge of the dynamics of the system, it is required that the variables to be estimated satisfy properties of algebraic observability.

The problem of anti-synchronization is another phenomenon of interest that occurs in chaotic oscillators. This problem has appeared in modern repetitions of Huygens' experiments [9], lasers [46, 47], salt-water oscillators [35] and some biological systems where a nonchaotic signal is generated [24]. Anti-synchronization has been treated as a direct modification of synchronization, simply with a sign change in the condition required for the error, and has been attacked with methods such as the active control [15, 20], the sliding mode control [11] and the sampled-data feedback L_∞ control [44].

Recently, systems with fractional dynamics, i.e. systems whose mathematical model is represented with derivatives and integrals of non-integer order, have been of great interest. This is mainly given to their applications to interdisciplinary areas such as material science [12], electromagnetism [42], electromechanics [48] and thermal systems [17]; in addition, in certain cases fractional equations give better approximations of the behaviour of the systems than the integer ones. Furthermore, it has been found that these systems can present chaotic behaviour, and also some strategies to control or synchronize them have been developed [1, 5–7, 22, 39–41].

7.2 Fractional Calculus Tools

In this section some definitions from fractional calculus that will be used in this chapter are defined briefly, as well as some existing stability results for fractional dynamical systems.

Definition 7.1 The Riemann–Liouville fractional integral of order $\alpha > 0$ of a function $f(t)$ is defined as

$$_0I_t^\alpha f(t) := \frac{1}{\Gamma(\alpha)} \int_0^t (t-\tau)^{\alpha-1} f(\tau) d\tau$$

where $\Gamma(\cdot)$ is Euler's Gamma function. The operator will be denoted as I^α in order to simplify the notation.

There are several definitions of fractional derivatives; the following is used in this chapter.

Definition 7.2 The Caputo fractional derivative of order $\alpha > 0$ of a function $f(t)$ is defined as

$$_0^C D_t^\alpha f(t) := I^{m-\alpha} D^m f(t) = \frac{1}{\Gamma(m-\alpha)} \int_0^t \frac{f^{(m)}(\tau)}{(t-\tau)^{\alpha-m+1}} d\tau$$

where $m - 1 < \alpha < m$, $m \in \mathbb{N}$, and $f^{(m)}(t)$ is the classical mth derivative of $f(t)$. The operator will be denoted as D^α in order to simplify the notation.

Now, consider the following class of fractional nonlinear systems:

$$D^\alpha x(t) = f(x) \tag{7.1}$$
$$y(t) = h(x)$$

where $x \in \mathbb{R}^n$ is the state vector, $y \in \mathbb{R}^p$ is the available output vector, $\boldsymbol{\alpha} = [\alpha_1, \alpha_2, \ldots, \alpha_n]^T$ is the vector of fractional orders, and f and h are analytic functions.

Definition 7.3 Consider the class of fractional nonlinear systems (7.1). If $\alpha_1 = \alpha_2 = \cdots = \alpha_n = \alpha \in \mathbb{R}$, system (7.1) is called a commensurate-order system, otherwise it is an incommensurate-order system.

In particular, $0 < \alpha_i < 1$ will be considered.

Definition 7.4 ([13]) Let $n_1, n_2 > 0$. The function $\mathscr{E}_{n_1,n_2}$ defined by

$$\mathscr{E}_{n_1,n_2}(z) = \sum_{k=0}^\infty \frac{z^k}{\Gamma(kn_1 + n_2)} \tag{7.2}$$

whenever the series converges is called the *two-parameter Mittag-Leffler function* with parameters n_1 and n_2.

Definition 7.5 ([26]) The solution of system $D^\alpha x(t) = f(t, x)$ is said to be Mittag-Leffler stable if

$$\|x(t)\| \leq \{m[x(0)]\mathcal{E}_{\alpha,1}(-\lambda t^\alpha)\}^b$$

$\alpha \in (0, 1)$, $\lambda \geq 0$, $b > 0$, $m(0) = 0$, $m(x) \geq 0$, and $m(x)$ is locally Lipschitz (with Lipschitz constant m_0) on $x \in \mathbb{B}$, an open subset of $\mathbb{R}^n$.

Theorem 7.1 ([26]) *Let $x = 0$ be an equilibrium point for the system $D^\alpha x(t) = f(t, x)$ and $\mathbb{D} \subset \mathbb{R}^n$ be a domain containing the origin. Let $V(t, x(t)) : [0, \infty) \times \mathbb{D} \to \mathbb{R}$ be a continuously differentiable function and locally Lipschitz with respect to x such that*

$$\alpha_1 \|x\|^a \leq V(t, x(t)) \leq \alpha_2 \|x\|^{ab}$$
$$D^\beta V(t, x(t)) \leq -\alpha_3 \|x\|^{ab}$$

where $t \geq 0$, $x \in \mathbb{D}$, $\beta \in (0, 1)$, α_1, α_2, α_3, a and b arbitrary positive constants. Then $x = 0$ is Mittag-Leffler stable. If the assumptions hold globally on $\mathbb{R}^n$, then $x = 0$ is globally Mittag-Leffler stable.

Lemma 7.1 ([2]) *Let $x(t) \in \mathbb{R}$ be a continuous and derivable function. Then, for any time instant $t \geq t_0$*

$$\frac{1}{2} D^\alpha x^2(t) \leq x(t) D^\alpha x(t) \qquad \forall \alpha \in (0, 1).$$

7.3　Fractional Synchronization and Anti-synchronization Problems

In this section, the problems of synchronization and anti-synchronization for fractional chaotic systems are defined, for both commensurate and incommensurate-order systems. The methodology via fractional integral reduced-order observer is explained.

7.3.1　Fractional Algebraic Observability Condition

Firstly, it is presented a condition that have to satisfy the non-measurable states of the chaotic systems in order to be able to reconstruct their dynamics. For commensurate systems, the following definition applies.

Definition 7.6 A state variable $\eta_i \in \mathbb{R}$ satisfies the fractional algebraic observability (FAO) condition if it is a function of the first $r \in \mathbb{N}$ sequential fractional derivatives of the available output y, i.e.,

$$\eta_i = \phi_i(y, D^\alpha y, D^{2\alpha} y, \ldots, D^{r\alpha} y) \tag{7.3}$$

where $\phi_i : \mathbb{R}^{(r+1)p} \to \mathbb{R}$.

Example 7.1 Classical Chua's oscillator is a simple electronic circuit that exhibits nonlinear dynamical phenomena such as bifurcation and chaos. The model in state equations is given by:

$$\dot{x}_1 = \rho(x_2 - x_1 - f(x))$$
$$\dot{x}_2 = x_1 - x_2 + x_3$$
$$\dot{x}_3 = -\beta x_2 - \gamma x_3$$

where $f(x) = m_1 x_1 + \frac{1}{2}(m_0 - m_1) \times (|x_1 + 1| - |x_1 - 1|)$, and $\rho, \beta, \gamma, m_0, m_1$ are parameters obtained from the values of the resistances, capacitances and inductances of the circuit.

The Chua–Hartley's system is different from Chua's system in that the piecewise-linear nonlinearity is replaced by an appropriate cubic nonlinearity which yields very similar behaviour. Derivatives on the left side of the differential equations are also replaced by the fractional derivatives [21].

Let the Chua–Hartley fractional commensurate oscillator be given by:

$$D^\alpha x_1 = \rho \left(x_2 + \frac{x_1 - 2x_1^3}{7} \right)$$
$$D^\alpha x_2 = x_1 - x_2 + x_3$$
$$D^\alpha x_3 = -\beta x_2$$

Taking the measurable output as $y = x_3$, the following relations can be obtained:

$$x_1 = \phi_1 \left(y, D^\alpha y, D^{2\alpha} y \right) = -\frac{1}{\beta} D^{2\alpha} y - \frac{1}{\beta} D^\alpha y - y$$

$$x_2 = \phi_2 \left(D^\alpha y \right) = -\frac{1}{\beta} D^\alpha y$$

and therefore, states x_1 and x_2 satisfy the FAO condition.

For incommensurate systems, the following definition is used.

Definition 7.7 A state variable $\eta_i \in \mathbb{R}$ satisfies the incommensurate fractional algebraic observability (IFAO) condition if it is a function of the derivatives of the available output y, i.e.,

$$\eta_i = \phi_i \left(y, y^{(\alpha_1)}, y^{(\alpha_1 + \alpha_2)}, \ldots, y^{\left(\sum_{i=1}^{n} c_i \alpha_i \right)} \right) \tag{7.4}$$

where $\phi_i : \mathbb{R}^{(r+1)p} \to \mathbb{R}$, $c_i \in \mathbb{N}$ and $0 \le \sum_{i=1}^{n} c_i \alpha_i \le 2$.

Example 7.2 Consider now the Chua–Hartley fractional incommensurate oscillator

$$D^{\alpha_1} x_1 = \rho \left(x_2 + \frac{x_1 - 2x_1^3}{7} \right)$$

$$D^{\alpha_2} x_2 = x_1 - x_2 + x_3$$

$$D^{\alpha_3} x_3 = -\beta x_2$$

Taking the measurable output as $y = x_1$, the following relations can be obtained:

$$x_2 = \phi_1 \left(y, D^{\alpha_1} y \right) = \frac{1}{\rho} D^{\alpha_1} y - \frac{1}{7} (y - 2y^3)$$

$$x_3 = \phi_2 \left(y, D^{\alpha_1} y, D^{\alpha_2} y, D^{\alpha_1 + \alpha_2} y \right) = \frac{1}{\rho} D^{\alpha_1 + \alpha_2} y$$

$$- \frac{1}{7} D^{\alpha_2} (y - 2y^3) - y + \frac{1}{\rho} D^{\alpha_1} y - \frac{1}{7} (y - 2y^3)$$

and therefore, states x_2 and x_3 satisfy the IFAO condition.

Remark 7.1 Each state that satisfies the FAO (or IFAO) condition is said to be algebraically observable, and thus its dynamics can be reconstructed.

7.3.2 Fractional Integral Reduced-Order Observer

Once determining that all the unknown states of the system are observable, the fractional synchronization problem can be solved by means of the master-slave configuration. Consider again system (7.1):

$$D^\alpha x(t) = f(x)$$

$$y(t) = h(x)$$

This dynamics can be extended to include an unknown vector state in a new variable η with unknown dynamics:

$$D^\alpha x = f(x, \eta)$$

$$D^\alpha \eta = \Delta(x)$$

$$y = h(x)$$

The problem is to construct the state η in order to determine the value of the desired state. But, given that the model has an unknown part, a fractional classical Luenberguer observer cannot be constructed.

In order to reconstruct the faults, a fractional-order reduced-order observer (ROO) will be used. The following is the fractional-order extension of a classic proportional

ROO (PROO):

$$D^\alpha \hat{f}_i = k_i(f_i - \hat{f}_i), \qquad 1 \le i \le q.$$

Moreover, the following is a fractional ROO which, besides the proportional part, adds fractional integral correction terms in order to improve its convergence. Thus this is a fractional integral reduced-order observer (FIROO):

$$D^\alpha \hat{\eta}_i = K_{i0}(\eta_i - \hat{\eta}_i) + K_{i1} I^\alpha(\eta_i - \hat{\eta}_i). \tag{7.5}$$

Remark 7.2 As its integer-order equivalent, the FIROO is model-free, i.e. it does not require to know the dynamics of the states, using only the FAO (or IFAO) condition to reconstruct them. Thus, it has an advantage against other methods and observers used for synchronization that require full knowledge of the system dynamics.

In order to work with the FIROO, it is assumed that the following hypotheses are satisfied:

H1 η_i satisfies the FAO (or IFAO) condition
H2 Let an auxiliary variable γ_i be a C^1 real-valued function.
H3 $\Delta_i(x)$ is bounded, i.e. $\exists N_i \in \mathbb{R}^+$ such that $\|\Delta_i(x)\| \le N_i, \forall x \in \Omega \subset \mathbb{R}^n$.

Theorem 7.2 *The FIROO (7.5) is Mittag-Leffler stable.*

Proof Define the observer error as $e_i = \eta_i - \hat{\eta}_i$. Consider the following Lyapunov function candidate:

$$V(e_i) = e_i^2 \tag{7.6}$$

Note that $V(e_i)$ satisfies the first inequality of Theorem 7.1 since:

$$\alpha_1 \|e_i\| \le V(e_i) \le \alpha_2 \|e_i\| \tag{7.7}$$

with $a = b = \alpha_1 = 1$ and $\alpha_2 = \sup(\|e_i\|)$.
By Lemma 7.1, it follows that

$$D^\alpha V(e_i) = D^\alpha e_i^2 \le 2e_i D^\alpha e_i$$

Therefore:

$$\begin{aligned}
D^\alpha V(e_i) &\le 2e_i D^\alpha e_i = 2e_i D^\alpha(\eta_i - \hat{\eta}_i) \\
&= 2e_i(\Delta_i - K_{i0}e_i - K_{i1} I^\alpha e_i) \\
&\le 2e_i \Delta_i - 2K_{i1}e_i I^\alpha e_i \\
&\le 2N_i \|e_i\| - 2K_{i1}\|e_i\|\|I^\alpha e_i\| \\
&\le -(2K_{i1}|I^\alpha e_i| - 2N_i)\|e_i\|
\end{aligned}$$

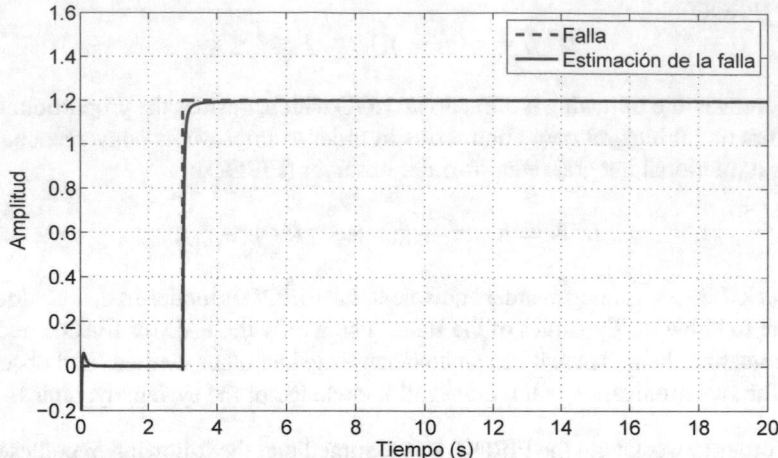

Fig. 7.1 Fault estimation by means of the PIROO

Then

$$D^\alpha V(e_i) \le -\alpha_3 \|e_i\| \tag{7.8}$$

with $\alpha_3 = 2K_{i1}|I^\alpha e_i| - 2N_i$.

Therefore, if $\alpha_3 > 0$, from Theorem 7.1 and Eqs. (7.6)–(7.8), it is concluded that the origin of system (7.5) is Mittag-Leffler stable.   □

Figures 7.1 and 7.2 show a comparison between the estimation of a step fault made with the PIROO and the estimation of the same fault made by the FIROO. It can be seen that, though it has a bigger overshoot, the FIROO improves convergence of the estimation.

7.3.3 Fractional Synchronization Problem

In this section, the fractional synchronization problem is solved as follows. The original fractional chaotic system will be known as the "master", because it acts as a driving system by means of its measurable output. Then, the FIROO will use this output in order to synchronize the dynamics of the estimated states with those of the master system; hence, the FIROO is known as the "slave" system. This problem is usually determined by the analysis of the fractional synchronization error dynamics, which is defined as follows:

$$e_i = \eta_i - \hat{\eta}_i \tag{7.9}$$

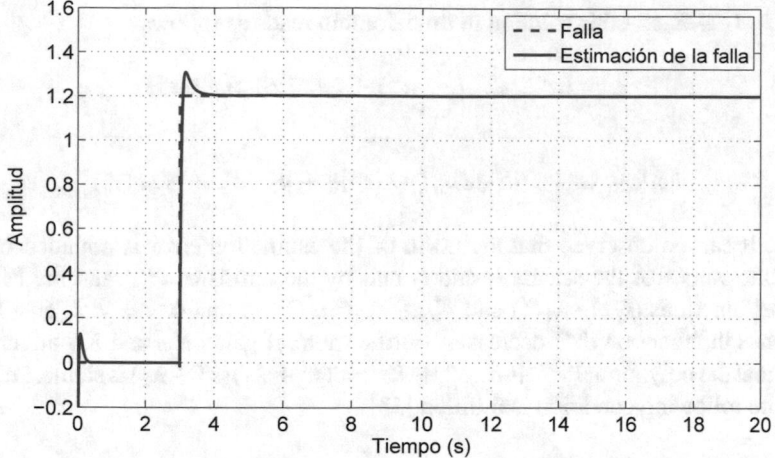

Fig. 7.2 Fault estimation by means of the FIROO

Considering the fractional dynamics of this error, the following is obtained:

$$D^\alpha e_i = D^\alpha \eta_i - D^\alpha \hat{\eta}_i$$
$$= D^\alpha \eta_i - K_{i0}e_i - K_{i1}I^\alpha e_i$$
$$D^\alpha e_i + K_{i0}e_i + K_{i1}I^\alpha e_i = D^\alpha \eta_i$$

The last equation is transformed to the Laplace domain:

$$s^\alpha E_i(s) - s^{\alpha-1}e_i(0) + K_{i0}E_i(s) + K_{i1}s^{-\alpha}E_i(s)$$
$$= s^\alpha H_i(s) - s^{\alpha-1}\eta_i(0)$$

from where the following solution is obtained:

$$E_i(s) = \frac{s^\alpha H_i(s) - s^{\alpha-1}\eta_i(0) + s^{\alpha-1}e_i(0)}{s^\alpha + K_{i0} + K_{i1}s^{-\alpha}}$$
$$= \frac{s^\alpha(s^\alpha H_i(s) - s^{\alpha-1}\eta_i(0) + s^{\alpha-1}e_i(0))}{s^{2\alpha} + K_{i0}s^\alpha + K_{i1}}$$

Note that this equation can be rewritten as:

$$E_i(s) = \frac{s^\alpha(s^\alpha H_i(s) - s^{\alpha-1}\eta_i(0) + s^{\alpha-1}e_i(0))}{(s^\alpha + \lambda_1)(s^\alpha + \lambda_2)}$$
$$= \frac{l_1}{s^\alpha + \lambda_1} + \frac{l_2}{s^\alpha + \lambda_2}$$

with $l_1, l_2 \in \mathbb{R}$, and this solution in time domain reads as follows:

$$e_i = l_1 t^{\alpha-1} \mathscr{E}_{\alpha,\alpha}(-\lambda_1 t^\alpha) + l_2 t^{\alpha-1} \mathscr{E}_{\alpha,\alpha}(-\lambda_2 t^\alpha)$$

So,

$$\|e_i\| \leq \|t^{\alpha-1}\|(|l_1|\|\mathscr{E}_{\alpha,\alpha}(-\lambda_1 t^\alpha)\| + |l_2|\|\mathscr{E}_{\alpha,\alpha}(-\lambda_2 t^\alpha)\|)$$

hence it can be observed that the norm of the estimation error is bounded by the absolute values of the scalars l_1 and l_2 and by the norms of $t^{\alpha-1}$ and the Mittag-Leffler functions $\mathscr{E}_{\alpha,\alpha}(-\lambda_1 t^\alpha)$ and $\mathscr{E}_{\alpha,\alpha}(-\lambda_2 t^\alpha)$. Given that $0 < \alpha < 1$, $\alpha - 1 < 0$ and thus the function $t^{\alpha-1}$ decreases. Furthermore, if gains K_{i0} and K_{i1} are chosen such that the polynomial $s^{2\alpha} + K_{i0}s^\alpha + K_{i1} = (s^\alpha + \lambda_1)(s^\alpha + \lambda_2)$ is stable, i.e. such that the following condition is fulfilled [33]:

$$|arg(\lambda_i)| > \alpha \frac{\pi}{2}, \quad i = 1, 2$$

then the polynomial is stable, and the Mittag-Leffler functions $\mathscr{E}_{\alpha,\alpha}(-\lambda_1 t^\alpha)$ and $\mathscr{E}_{\alpha,\alpha}(-\lambda_2 t^\alpha)$ tend to the origin of the error dynamics, which makes $\hat{\eta}_i$ to follow η_i. Thus, the fractional synchronization problem is solved.

7.3.4 Fractional Anti-synchronization Problem

The problem of fractional anti-synchronization consists in obtaining also an estimation of the unknown states of the master system, but instead of having the state of the slave system converge to the value of the state of the master, it will converge to that value with opposite sign. For this, a new variable $\hat{\xi}_i = -\hat{\eta}_i$ is introduced, which is the estimated state of the anti-synchronized system. So, the fractional anti-synchronization error dynamics is defined as follows:

$$\bar{e}_i = \eta_i + \hat{\xi}_i \tag{7.10}$$

In a procedure similar to the former, the following inequality is obtained:

$$\|\bar{e}_i\| \leq \|t^{\alpha-1}\|(|l_1|\|\mathscr{E}_{\alpha,\alpha}(-\lambda_1 t^\alpha)\| + |l_2|\|\mathscr{E}_{\alpha,\alpha}(-\lambda_2 t^\alpha)\|)$$

hence selecting the appropriate values of gains K_{i0} and K_{i1}, the polynomial $s^{2\alpha} + K_{i0}s^\alpha + K_{i1} = (s^\alpha + \lambda_1)(s^\alpha + \lambda_2)$ is stable, and the Mittag-Leffler functions $\mathscr{E}_{\alpha,\alpha}(-\lambda_1 t^\alpha)$ and $\mathscr{E}_{\alpha,\alpha}(-\lambda_2 t^\alpha)$ tend to the origin of the error dynamics, which makes $\hat{\xi}_i$ to follow η_i but with an inverse sign. Thus, the fractional anti-synchronization problem is solved.

Remark 7.3 This methodology works for commensurate and incommensurate-order systems, given that the respective unknown variables satisfy the FAO or IFAO property, respectively.

7.4 Application to Fractional Chaotic Systems

In this section, the FIROO is used to achieve synchronization and anti-synchronization in two fractional chaotic systems, namely the Lorenz and the Rössler oscillators. The former will be treated as a commensurate-order system, and the latter as an incommensurate-order one.

7.4.1 Fractional Lorenz System

The Lorenz oscillator is an attractor named after Edward M. Lorenz, who derived it from the simplified equations of turbulent convection rolls arising in the equations of the atmosphere [27]. This system is related to the so-called "butterfly effect". The generalization to fractional dynamics was presented in [18].

Consider the commensurate fractional-order chaotic Lorenz system:

$$D^\alpha x_1 = \sigma(x_2 - x_1) \tag{7.11}$$
$$D^\alpha x_2 = \rho x_1 - x_2 - x_1 x_3$$
$$D^\alpha x_3 = x_1 x_2 - \beta x_3$$
$$y = x_1$$

with $\sigma, \rho, \beta > 0$.

Firstly, it has to be verified that the unknown states, x_2 and x_3, satisfy the FAO condition. From (7.11) the following equations can be obtained:

$$x_2 = \phi(y, D^\alpha y) = (1/\sigma)(\sigma y + D^\alpha y) \tag{7.12}$$
$$x_3 = \phi(y, D^\alpha y, D^{2\alpha} y)$$
$$= \rho - 1 - (1/\sigma y)(D^\alpha y(\sigma + 1) + D^{2\alpha} y)$$

and thus, both states are algebraically observable, so the FIROO can be built to reconstruct them.

Let $\eta_2 = x_2$. Using (7.12), the FIROO for this variable is:

$$D^\alpha \hat{\eta}_2 = K_{10}(\eta_2 - \hat{\eta}_2) + K_{11} I^\alpha (\eta_2 - \hat{\eta}_2)$$
$$= K_{10}\left(\frac{1}{\sigma}(\sigma y + D^\alpha y) - \hat{\eta}_2\right)$$
$$+ K_{11} I^\alpha \left(\frac{1}{\sigma}(\sigma y + D^\alpha y) - \hat{\eta}_2\right)$$

Define the auxiliary variable $\gamma_1 = \hat{\eta}_2 - \frac{K_{10}}{\sigma} y$. Then:

$$D^\alpha \gamma_1 = D^\alpha \hat{\eta}_2 - \frac{K_{10}}{\sigma} D^\alpha y$$

$$= K_{10}(y - \hat{\eta}_2) + \frac{K_{11}}{\sigma} y + K_{11} I^\alpha (y - \hat{\eta}_2)$$

So, the FIROO for $\hat{\eta}_2$ is:

$$D^\alpha \gamma_1 = K_{10}\left(y - \gamma_1 - \frac{K_{10}}{\sigma} y\right) + \frac{K_{11}}{\sigma} y \tag{7.13}$$

$$+ K_{11} I^\alpha \left(y - \gamma_1 - \frac{K_{10}}{\sigma} y\right)$$

$$\hat{\eta}_2 = \gamma_1 + \frac{K_{10}}{\sigma} y \tag{7.14}$$

Now let $\eta_3 = x_3$. Note that from (7.11), the following relation can be obtained:

$$x_3 = (1/\beta)[(y/\sigma)(\sigma y + D^\alpha y) - D^\alpha \eta_3] \tag{7.15}$$

Using this equation, the FIROO for this variable is:

$$D^\alpha \hat{\eta}_3 = K_{20}(\eta_3 - \hat{\eta}_3) + K_{21} I^\alpha (\eta_3 - \hat{\eta}_3)$$

$$= K_{20}\left(\frac{1}{\beta}\left(\frac{1}{\sigma} y(\sigma y + D^\alpha y) - D^\alpha \hat{\eta}_3\right) - \hat{\eta}_3\right)$$

$$+ K_{21} I^\alpha \left(\frac{1}{\beta}\left(\frac{1}{\sigma} y(\sigma y + D^\alpha y) - D^\alpha \hat{\eta}_3\right) - \hat{\eta}_3\right)$$

After some algebraic manipulations, the following is obtained:

$$D^\alpha \hat{\eta}_3 = \frac{K_{20}}{(\beta + K_{20})\sigma} y D^\alpha y$$

$$+ \frac{K_{20}\beta}{\beta + K_{20}}\left[\left(\frac{1}{\beta} - \frac{K_{20}}{2(\beta + K_{20})\sigma}\right) y^2 - \hat{\eta}_3\right.$$

$$+ \left.\frac{K_{20}}{2(\beta + K_{20})\sigma} y^2\right] - \frac{K_{20}K_{21}}{\beta + K_{20}} \hat{\eta}_3$$

$$+ \frac{K_{20}K_{21}\beta}{\beta + K_{20}} I^\alpha \left(\frac{1}{\beta\sigma} y(\sigma y + D^\alpha y) - \hat{\eta}_3\right)$$

Define the auxiliary variable $\gamma_2 = \hat{\eta}_3 - \frac{K_{20}}{2(\beta + K_{20})\sigma} y^2$. Then:

$$D^{\alpha}\gamma_2 = D^{\alpha}\hat{\eta}_3 - \frac{K_{20}}{(\beta + K_{20})\sigma}yD^{\alpha}y$$

$$= \frac{K_{20}\beta}{\beta + K_{20}}\left[\left(\frac{1}{\beta} - \frac{K_{20}}{2(\beta + K_{20})\sigma}\right)y^2 - \gamma_2\right]$$

$$- \frac{K_{20}K_{21}}{\beta + K_{20}}\hat{\eta}_3$$

$$+ \frac{K_{20}K_{21}\beta}{\beta + K_{20}}I^{\alpha}\left(\frac{1}{\beta\sigma}y(\sigma y + D^{\alpha}y) - \hat{\eta}_3\right)$$

So, the FIROO for $\hat{\eta}_3$ is:

$$D^{\alpha}\gamma_2 = \frac{K_{20}\beta}{\beta + K_{20}}\left[\left(\frac{1}{\beta} - \frac{K_{20}}{2(\beta + K_{20})\sigma}\right)y^2 - \gamma_2\right] \qquad (7.16)$$

$$- \frac{K_{20}K_{21}}{\beta + K_{20}}\left(\gamma_2 + \frac{K_{20}}{2(\beta + K_{20})\sigma}y^2\right)$$

$$+ \frac{K_{20}K_{21}\beta}{\beta + K_{20}}I^{\alpha}\left(\frac{1}{\beta\sigma}y(\sigma y + D^{\alpha}y)\right.$$

$$\left. - \gamma_2 - \frac{K_{20}}{2(\beta + K_{20})\sigma}y^2\right)$$

$$\hat{\eta}_3 = \gamma_2 + \frac{K_{20}}{2(\beta + K_{20})\sigma}y^2 \qquad (7.17)$$

Thus, system (7.11) acts as the master system and systems (7.13) and (7.16) form the slave system. For the synchronization problem, the unknown states are obtained with (7.14) and (7.17). For the anti-synchronization problem, the same equations are used, but defining the states $\hat{\xi}_2 = -\hat{\eta}_2$ and $\hat{\xi}_3 = -\hat{\eta}_3$.

Simulations were performed with the Ninteger Toolbox from Matlab-Simulink® during 20 s, with $\alpha = 0.993$, parameters $\sigma = 10$, $\rho = 28$, $\beta = 8/3$ and gains $K_{10} = 250$, $K_{11} = 1$, $K_{20} = 150$ and $K_{21} = 0.01$. The initial conditions were set as $x_1(0) = x_2(0) = x_3(0) = 10$, $\gamma_2(0) = \gamma_3(0) = -10$.

Figures 7.3 and 7.4 show the synchronization between the states x_2 and x_3 of the master and their estimations $\hat{\eta}_2$ and $\hat{\eta}_3$, respectively, from the slave. Figures 7.5 and 7.6 show the anti-synchronization between the states x_2 and x_3 of the master and their estimations $\hat{\xi}_2$ and $\hat{\xi}_3$, respectively, from the slave. Finally, Fig. 7.7 shows the synchronization and anti-synchronization signals in state space.

7.4.2 Fractional Rössler System

This attractor was designed Otto Rössler in 1976; its originally theoretical equations were later found to be useful in modeling equilibrium in chemical reactions [43]. The generalization to fractional dynamics was proposed in [25].

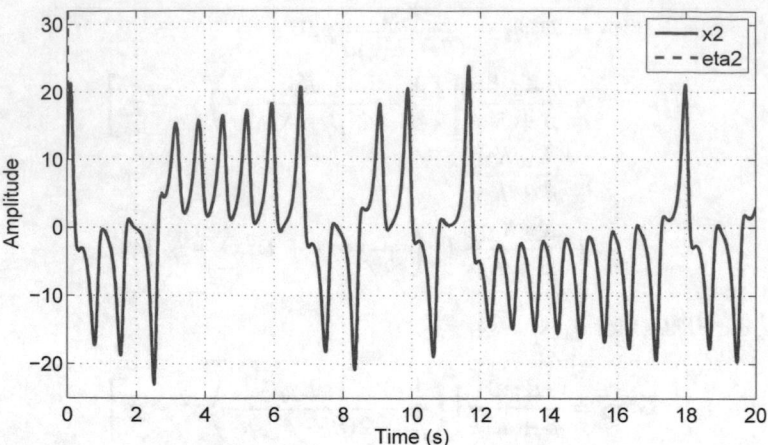

Fig. 7.3 Synchronization between x_2 and $\hat{\eta}_2$

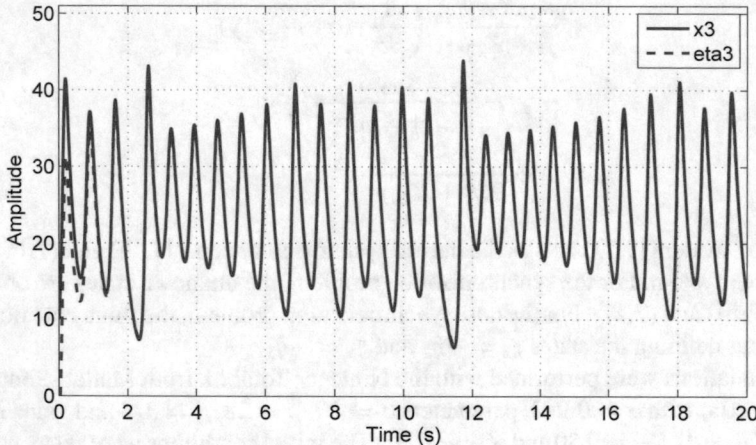

Fig. 7.4 Synchronization between x_3 and $\hat{\eta}_3$

Consider the incommensurate fractional-order chaotic Rössler system:

$$D^{\alpha_1} x_1 = -x_2 - x_3 \tag{7.18}$$
$$D^{\alpha_2} x_2 = x_1 + a x_2$$
$$D^{\alpha_3} x_3 = 0.2 + x_3(x_1 - 10)$$
$$y = x_2$$

where a is allowed to be varied.

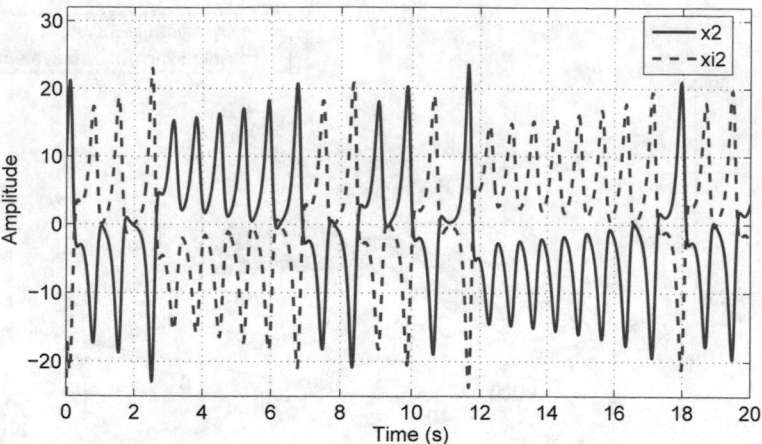

Fig. 7.5 Anti-synchronization between x_2 and $\hat{\xi}_2$

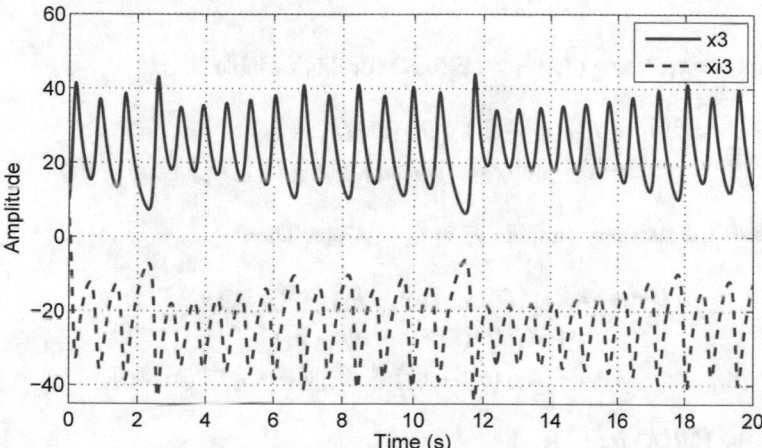

Fig. 7.6 Anti-synchronization between x_3 and $\hat{\xi}_3$

Firstly, it has to be verified that the unknown states, x_1 and x_3, satisfy the IFAO condition. From (7.18) the following equations can be obtained:

$$x_1 = \phi(y, D^{\alpha_2}y) = -ay + D^{\alpha_2}y \tag{7.19}$$

$$x_3 = \phi(y, D^{\alpha_1}y, D^{\alpha_1+\alpha_2}y) \tag{7.20}$$
$$= -y + aD^{\alpha_1}y - D^{\alpha_1+\alpha_2}y$$

and thus, both states are algebraically observable, so the FIROO can be built to reconstruct them.

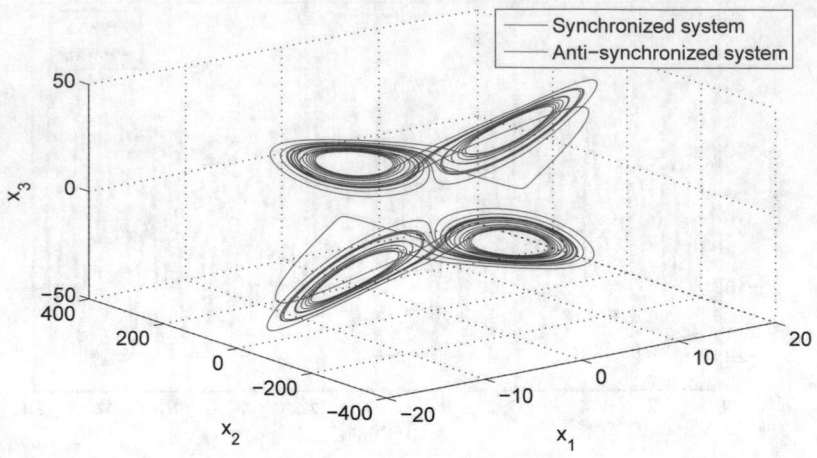

Fig. 7.7 Synchronization and anti-synchronization of the Lorenz system in state space

Let $\eta_1 = x_1$. Using (7.19), the FIROO for this variable is:

$$D^{\alpha_2}\hat{\eta}_1 = K_{10}(\eta_1 - \hat{\eta}_1) + K_{11}I^{\alpha_2}(\eta_1 - \hat{\eta}_1)$$
$$= K_{10}\left(D^{\alpha_2}y - ay - \hat{\eta}_1\right) + K_{11}I^{\alpha_2}(D^{\alpha_2}y - ay - \hat{\eta}_1)$$

Define the auxiliary variable $\gamma_1 = \hat{\eta}_1 - K_{10}y$. Then:

$$D^{\alpha_2}\gamma_1 = K_{10}\left(D^{\alpha_2}y - ay - \hat{\eta}_1\right) - K_{10}D^{\alpha_2}y$$
$$+ K_{11}I^{\alpha_2}(D^{\alpha_2}y - ay - \hat{\eta}_1)$$
$$= -K_{10}\left(ay + \hat{\eta}_1\right) + K_{11}y - K_{11}I^{\alpha_2}(ay + \hat{\eta}_1)$$

So, the FIROO for $\hat{\eta}_1$ is:

$$D^{\alpha_2}\gamma_1 = -K_{10}\left(ay + \gamma_1 + K_{10}y\right) + K_{11}y \tag{7.21}$$
$$- K_{11}I^{\alpha_2}(ay + \gamma_1 + K_{10}y)$$
$$\hat{\eta}_1 = \gamma_1 + K_{10}y \tag{7.22}$$

Let $\eta_3 = x_3$. Using (7.20), the FIROO for this variable is:

$$D^{\alpha_2}\hat{\eta}_3 = K_{20}(\eta_3 - \hat{\eta}_3) + K_{21}I^{\alpha_2}(\eta_3 - \hat{\eta}_3) \tag{7.23}$$
$$= K_{20}\left(-y - D^{\alpha_1+\alpha_2}y + aD^{\alpha_1}y - \hat{\eta}_3\right)$$
$$+ K_{21}I^{\alpha_2}\left(-y - D^{\alpha_1+\alpha_2}y + aD^{\alpha_1}y - \hat{\eta}_3\right)$$

Note that this equation deals with both $D^{\alpha_1}y$ and $D^{\alpha_2}y$, so a variable $\zeta = D^{\alpha_1}y$ is defined, and is also estimated via FIROO:

$$D^{\alpha_1}\hat{\zeta} = K_{\zeta 0}(\zeta - \hat{\zeta}) + K_{\zeta 1}I^{\alpha_1}(\zeta - \hat{\zeta})$$
$$= K_{\zeta 0}(D^{\alpha_1}y - \hat{\zeta}) + K_{\zeta 1}I^{\alpha_1}(D^{\alpha_1}y - \hat{\zeta})$$

Define the auxiliary variable $\gamma_\zeta = \hat{\zeta} - K_{\zeta 0}y$. Then:

$$D^{\alpha_1}\gamma_\zeta = K_{\zeta 0}(D^{\alpha_1}y - \hat{\zeta}) - K_{\zeta 0}D^{\alpha_1}y$$
$$+ K_{\zeta 1}I^{\alpha_1}(D^{\alpha_1}y - \hat{\zeta})$$
$$= -K_{\zeta 0}\hat{\zeta} + K_{\zeta 1}y - K_{\zeta 1}I^{\alpha_1}\hat{\zeta}$$

So, the FIROO for $\hat{\zeta}$ is:

$$D^{\alpha_1}\gamma_\zeta = -K_{\zeta 0}\left(\gamma_\zeta + K_{\zeta 0}y\right) + K_{\zeta 1}y \tag{7.24}$$
$$- K_{\zeta 1}I^{\alpha_1}\left(\gamma_\zeta + K_{\zeta 0}y\right)$$
$$\hat{\zeta} = \gamma_\zeta + K_{\zeta 0}y \tag{7.25}$$

Substitution of (7.25) into (7.23) leads to:

$$D^{\alpha_2}\hat{\eta}_3 = K_{20}(-y - D^{\alpha_2}\hat{\zeta} + a\hat{\zeta} - \hat{\eta}_3)$$
$$+ K_{21}I^{\alpha_2}(-y - D^{\alpha_2}\hat{\zeta} + a\hat{\zeta} - \hat{\eta}_3)$$

Define the auxiliary variable $\gamma_2 = \hat{\eta}_3 + K_{20}\hat{\zeta}$. Then:

$$D^{\alpha_2}\gamma_2 = K_{20}(-y - D^{\alpha_2}\hat{\zeta} + a\hat{\zeta} - \hat{\eta}_3) + K_{20}D^{\alpha_2}\hat{\zeta}$$
$$+ K_{21}I^{\alpha_2}(-y - D^{\alpha_2}\hat{\zeta} + a\hat{\zeta} - \hat{\eta}_3)$$
$$= K_{20}(-y + a\hat{\zeta} - \hat{\eta}_3) - K_{21}\hat{\zeta}$$
$$+ K_{21}I^{\alpha_2}(-y + a\hat{\zeta} - \hat{\eta}_3)$$

Finally, the FIROO for $\hat{\eta}_3$ is:

$$D^{\alpha_2}\gamma_2 = -K_{20}(y - a\hat{\zeta} + \gamma_2 - K_{20}\hat{\zeta}) - K_{21}\hat{\zeta} \tag{7.26}$$
$$- K_{21}I^{\alpha_2}(y - a\hat{\zeta} + \gamma_2 - K_{20}\hat{\zeta})$$
$$\hat{\eta}_3 = \gamma_2 - K_{20}\hat{\zeta} \tag{7.27}$$

Thus, system (7.18) acts as the master system and systems (7.21) and (7.26) form the slave system. For the synchronization problem, the unknown states are obtained with (7.22) and (7.27). For the anti-synchronization problem, the same equations are used, but defining the states $\hat{\xi}_1 = -\hat{\eta}_1$ and $\hat{\xi}_3 = -\hat{\eta}_3$.

Simulations were performed during 100 s, with $\alpha_1 = 0.9$, $\alpha_2 = 0.8$, $\alpha_3 = 0.7$, parameter $a = 0.63$ and gains $K_{10} = 120$, $K_{11} = 1$, $K_{\zeta 0} = 100$, $K_{\zeta 1} = 1$, $K_{20} = 10$ and $K_{21} = 0.01$. The initial conditions were set as $x_1(0) = 1, x_2(0) = 0, x_3(0) = -5$, $\gamma_1(0) = 3$, $\gamma_\zeta(0) = \gamma_3(0) = 1$.

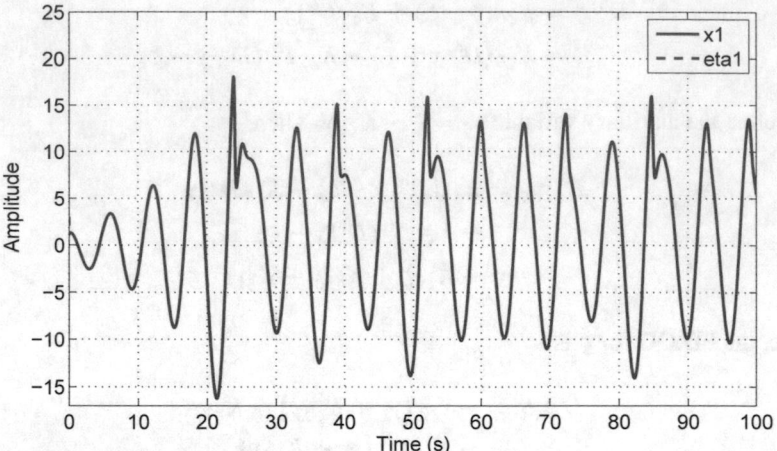

Fig. 7.8 Synchronization between x_1 and $\hat{\eta}_1$

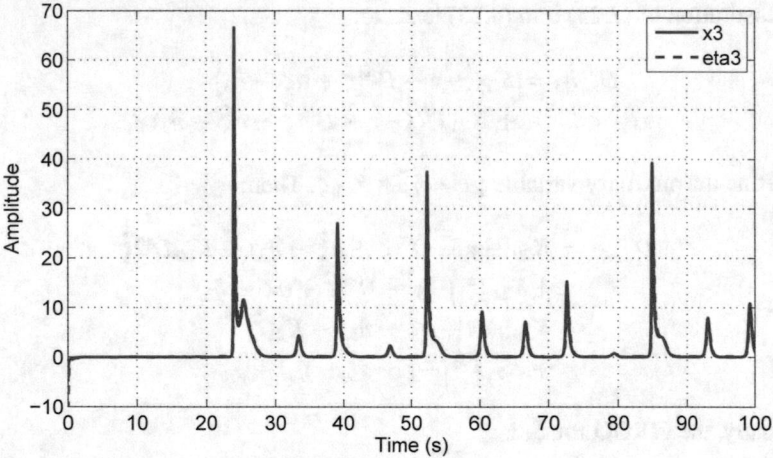

Fig. 7.9 Synchronization between x_3 and $\hat{\eta}_3$

Figures 7.8 and 7.9 show the synchronization between the states x_1 and x_3 of the master and their estimations $\hat{\eta}_1$ and $\hat{\eta}_3$, respectively, from the slave. Figures 7.10 and 7.11 show the anti-synchronization between the states x_1 and x_3 of the master and their estimations $\hat{\xi}_1$ and $\hat{\xi}_3$, respectively, from the slave. Finally, Fig. 7.12 shows the synchronization and anti-synchronization signals in state space.

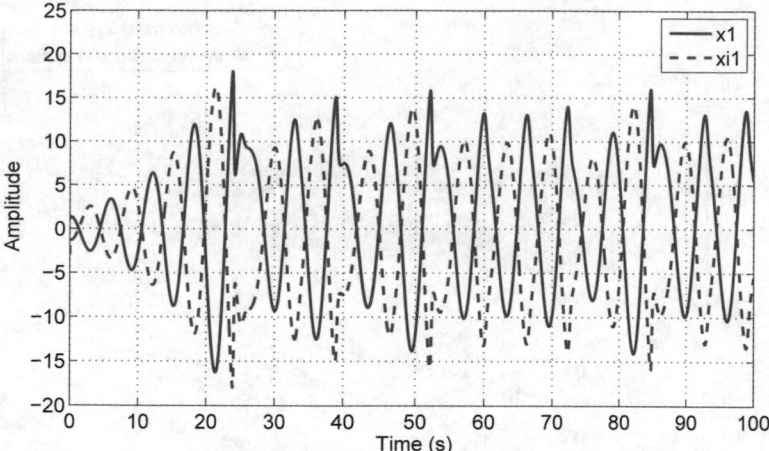

Fig. 7.10 Anti-synchronization between x_1 and $\hat{\xi}_1$

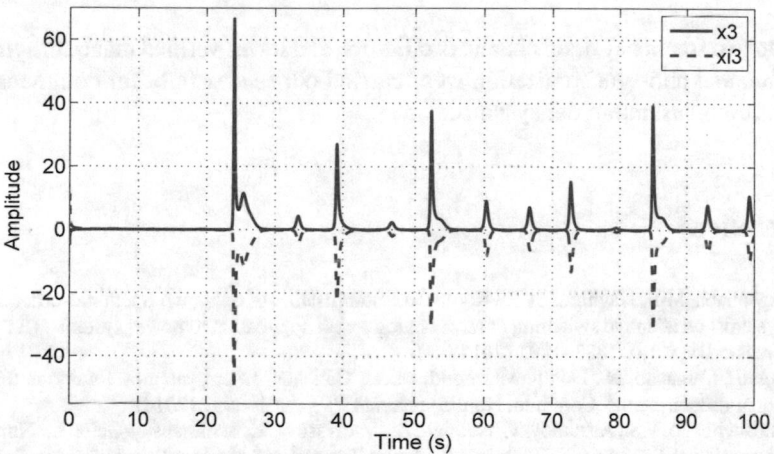

Fig. 7.11 Anti-synchronization between x_3 and $\hat{\xi}_3$

7.5 Concluding Remarks

In this chapter the FIROO was introduced, which was shown to be Mittag-Leffler stable, was proposed to solve the synchronization and anti-synchronization problems in commensurate and incommensurate-order fractional chaotic systems, by means of the master-slave configuration. The proposed observer served as a slave system while the chaotic system acted as the master. To be able to use this observer, the unknown variables of the master systems had to satisfy the fractional algebraic observability condition. Simulations were performed using the proposed methodol-

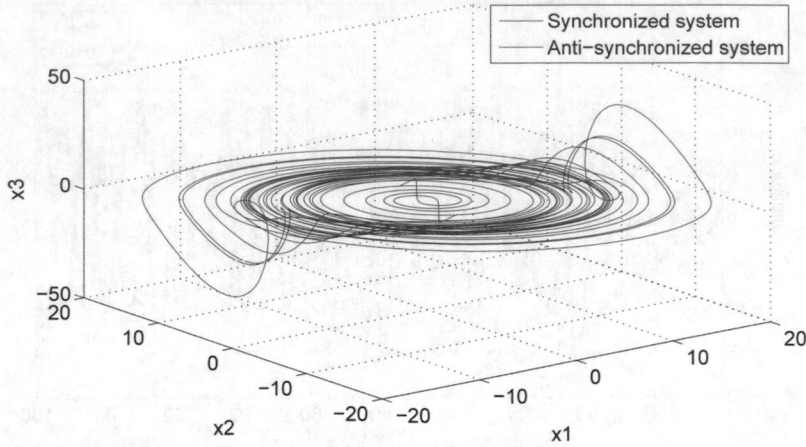

Fig. 7.12 Synchronization and anti-synchronization of the Rössler system in state space

ogy on two fractional-order chaotic oscillators, and it was verified that both synchronization and anti-synchronization were carried out successfully for commensurate and incommensurate-order systems.

References

1. Aghababa, M.P., Haghighi, A.R., Roohi, M.: Stabilisation of unknown fractional-order chaotic systems: an adaptive switching control strategy with application to power systems. IET Gener Transm Dis **9**(14), 1883–1893 (2015)
2. Aguila-Camacho, N., Duarte-Mermoud, M.A., Gallegos, J.A.: Lyapunov functions for fractional order systems. Commun. Nonlinear Sci. **19**(9), 2951–2957 (2014)
3. Anishchenko, V.S., Astakhov, V., Neiman, A., Vadivasova, T., Schimansky-Geier, L.: Nonlinear Dynamics of Chaotic and Stochastic Systems: Tutorial and Modern Developments. Springer, Berlin (2007)
4. Balanov, A., Janson, N., Postnov, D., Sosnovtseva, O.: Synchronization: From Simple to Complex. Springer, Berlin (2009)
5. Bao, H., Park, J.H., Cao, J.: Adaptive synchronization of fractional-order memristor-based neural networks with time delay. Nonlinear Dynam. **82**(3), 1343–1354 (2015)
6. Bao, H., Park, J.H., Cao, J.: Synchronization of fractional-order complex-valued neural networks with time delay. Neural Netw. **81**(2016), 16–28 (2016)
7. Bao, H., Park, J.H., Cao, J.: Synchronization of fractional-order delayed neural networks with hybrid coupling. Complexity **21**(2016), 106–112 (2016)
8. Baumann, M., Leine, R.I.: Synchronization-based state observer for impacting multibody systems using switched geometric unilateral constraints. In: Proceedings of the ECCOMAS Thematic Conference on Multibody Dynamics, June 29–July 2, pp. 585–596. Barcelona, Catalonia, Spain (2015)
9. Bennett, M., Schatz, M.F., Rockwood, H., Wiesenfeld, K.: Huygens's clocks. Proc. Math. Phys. Eng. Sci. **458**(2019), 563–579 (2012)
10. Canyelles-Pericas, P., Busawon, K.: High gain observer with algorithm transformation to extended Jordan observable form for chaos synchronization applications. In: Proceedings of

the: UKACC International Conference on Control, 9th–11th July 2014, pp. 262–267. Lough-borough, U.K. (2014)

11. Chen, D., Zhang, R., Ma, X., Liu, S.: Chaotic synchronization and anti-synchronization for a novel class of multiple chaotic systems via a sliding mode control scheme. Nonlinear Dynam. **69**(1–2), 35–55 (2012)

12. De Espíndola, J.J., da Silva Neto, J.M., Lopes, E.M.O.: A generalized fractional derivative approach to viscoelastic material properties measurement. Appl. Math. Comput. **164**(2), 493–506 (2005)

13. Diethelm, K.: The Analysis of Fractional Differential Equations: An Application-Oriented Exposition Using Differential Operators of Caputo Type. Springer, Berlin (2010)

14. Dimassi, H., Loria, A., Belghith, S.: Adaptive observers-based synchronization of a class of Lur'e systems with delayed outputs for chaotic communications. In: Proceedings of the: IFAC Conference on Analysis and Control of Chaotic Systems, 20–22 June 2012, pp. 255–260. Cancún, México (2012)

15. Emadzadeh, A.A., Haeri, M.: Anti-synchronization of two different chaotic systems via active control. Proc. Wrld Acad. Sci. E **6**(2005), 62–65 (2005)

16. Engbert, R., Drepper, F.R.: Chance and chaos in population biology - models of recurrent epidemics and food chain dynamics. Chaos Soliton Fract. **4**(1994), 1147–1169 (1994)

17. Gabano, J.D., Poinot, T.: Fractional modelling and identification of thermal systems. Signal Process **91**(3), 531–541 (2011)

18. Grigorenko, I., Grigorenko, E.: Chaotic dynamics of the fractional Lorenz system. Phys. Rev. Lett. **91**(3), 034101-1–034101-4 (2003)

19. Guckenheimer, J., Homes, P.: Nonlinear Oscillations, Dynamical Systems, and Bifurcations of Vector Fields. Springer, New York (1993)

20. Guo-Hui, L.: Synchronization and anti-synchronization of Colpitts oscillators using active control. Chaos Solitons Fract. **26**(1), 87–93 (2005)

21. Hartley, T.T., Lorenzo, C.F., Qammer, H.K.: Chaos in a fractional-order Chua's system. IEEE Trans. Circuits-I **42**(8), 485–490 (1995)

22. Huang, C., Cao, J.: Active control strategy for synchronization and anti-synchronization of a fractional chaotic financial system. Phys. A **473**(2017), 262–275 (2017)

23. Kaddoum, G., Coulon, M., Roviras, D., Chargé, P.: Theoretical performance for asynchronous multi-user chaos-based communication systems on fading channels. Signal Process **90**(11), 2923–2933 (2010)

24. Kim, C.M., Rim, S., Kye, W.H., Ryu, J.W., Park, Y.J.: Anti-synchronization of chaotic oscillators. Phys. Lett. A **320**(1), 39–46 (2003)

25. Li, C., Chen, G.: Chaos and hyperchaos in the fractional-order Rössler equations. Phys. A **341**(2004), 55–61 (2004)

26. Li, Y., Chen, Y., Podlubny, I.: Stability of fractional-order nonlinear dynamic systems: Lya-punov direct method and generalized Mittag-Leffler stability. Comput. Math. Appl. **59**(5), 1810–1821 (2010)

27. Lorenz, E.N.: Deterministic nonperiodic flow. J. Atmos. Sci. **20**(1963), 130–141 (1963)

28. Mata, J.L., Martínez-Guerra, R., Aguilar, R.: A new asymptotic polynomial observer to syn-chronization problem. In: Proceedings of the 6th International Conference on Electrical Engi-neering, Computing Science and Automatic Control (CCE 2009), 10–13 November 2009, pp. 66–71. Toluca, México (2009)

29. Mata, J.L., Martínez-Guerra, R., Aguilar, R.: A bounded error observer for synchronization of chaotic systems. In: Proceedings of the: 6th International Conference on Electrical Engineering, Computing Science and Automatic Control (CCE 2009), 10–13 November 2009, pp. 72–77. Toluca, México (2009)

30. Martínez-Guerra, R., Cruz-Ancona, C.D., Pérez-Pinacho, C.A.: Generalized multi-synchronization viewed as a multi-agent leader-following consensus problem. Appl. Math. Comput. **282**(2016), 226–236 (2016)

31. Martínez-Guerra, R., Luviano-Juárez, A., Rincón-Pasaye, J.J.: On nonlinear observers: a dif-ferential algebraic approach. In: Proceedings of the: American Control Conference, 11–13 July 2007, pp. 1682–1686. New York City, USA (2007)

32. Martínez-Guerra, R., Montesinos García, J.J., Delfín-Prieto, S.M.: Secure communications via synchronization of Liouvillian Chaotic systems. J. Frankl Inst. **353**(17), 4384–4399
33. Matignon, D.: Stability results for fractional differential equations with applications to control processing. Ma Comput. Sci. Eng. **2**(1996), 963–968 (1996)
34. Mosekilde, E., Maistrenko, Y., Postnov, D.: Chaotic Synchronization Application to Living Systems. World Scientific Publishing, Singapore (2002)
35. Nakata, S., Miyata, T., Ojima, N., Yoshikawa, K.: Self-synchronization in coupled salt-water oscillators. Phys. D **115**(1998), 313–320 (1998)
36. Pecora, L.M., Carroll, T.L.: Synchronization in Chaotic systems. Phys. Rev. Lett. **64**(8), 821–824 (1990)
37. Pikovsky, A., Rosenblum, M., Kurths, J.: Synchronization: A Universal Concept in Nonlinear Sciences. Cambridge University Press, Cambridge (2001)
38. Poinsard, S., Loría, A.: Robust Communication-Masking via a Synchronized Chaotic Lorenz Transmission System. In: Luo, A.C.J., Barbosa, R.S., Silva, M.F., Figueiredo, L.B. Tenreiro-Machado, J.A. (eds.) Nonlinear Science and Complexity, pp. 357–365. Springer, Berlin (2011)
39. Razminia, A., Baleanu, D.: Complete synchronization of commensurate fractional order chaotic systems using sliding mode control. Mechatronics **23**(7), 873–879 (2013)
40. Razminia, A., Majd, V.J., Baleanu, D.: Chaotic incommensurate fractional order Rössler system: active control and synchronization. Adv. Differ. Equ.-NY **2011**, 15 (2011)
41. Razminia, A., Torres, D.F.M.: Control of a novel chaotic fractional order system using a state feedback technique. Mechatronics **23**(7), 755–763 (2013)
42. Rosales, J.J., Gómez, J.F., Guía, M., Tkach, V.I.: Fractional electromagnetic waves. In: Proceedings of the LFNM*2011 International Conference on Laser & Fiber-Optical Networks Modeling, Kharkov, Ukraine, 4–8 September, pp. 1–3 (2011)
43. Rössler, O.E.: An equation for continuous chaos. Phys. Lett. A **57**(1976), 397–398 (1976)
44. Sakthivel, R., Santra, S., Anthoni, S.M., Kuppili, V.: Synchronisation and anti-synchronisation of chaotic systems with application to DC-DC boost converter. IET Gener. Transm. Dis. **11**(4), 959–967 (2017)
45. Solak, E.: A reduced-order observer for the synchronization of Lorenz systems. Phys. Lett. A **325**(2004), 276–278 (2004)
46. Uchida, A., Liu, Y., Fischer, I., Davis, P., Aida, T.: Chaotic antiphase dynamics and synchronization in multimode semiconductor lasers. Phys. Rev. A **64**, 023801-1–023801-6 (2001)
47. Wedekind, I., Parlitz, U.: Experimental observation of synchronization and anti-synchronization of chaotic low-frequency-fluctuations in external cavity semiconductor lasers. Int. J. Bifurcat. Chaos **11**(4), 1141–1147 (2001)
48. Yu, W., Luo, Y., Pi, Y.: Fractional order modeling and control for permanent magnet synchronous motor velocity servo system. Mechatronics **23**(7), 813–820 (2013)

Chapter 8
Multi-fault-tolerant Control in Fractional-Order Systems

Fractional-order dynamical systems, in contrast to the classical integer-order systems, have been strongly studied in the last decades. This is due to the great amount of applications and physical problems that present dynamics with fractional derivatives and integrals, such as materials science [10], thermal systems [12], diffusion problems [27], viscoelasticity [35], polymer behavior [23], finance [34], bioengineering [20], damped mechanical systems [14], electrical circuits [18], electromagnetism [32], electromechanics [44], etc. In particular, in control theory the generalization of the PID controller and other controllers that involve fractional dynamics have been developed, such as the CRONE [33] and the fractional sliding mode controller [25, 31]. Some recent specific studies regarding fractional order systems can be found, for example one that deals with the existence of solutions of a class of fractional stochastic differential equations with delays [4]; other paper proposes an identification algorithm for a time-delay fractional-order system with measurement noise [13]; other work deals with the analysis of bifurcations in a delayed fractional complex-valued neural network [16]; in another article the implementation of some classes of commensurate fractional-order transfer functions is performed with fractional-order capacitors [39]; other work investigates the problem of on line parameter estimation for fractional-order linear systems by making an extension of the gradient algorithm [41]; in another paper an analysis to some approximations of the fractional order Van der Pol Oscillator is performed [43].

Furthermore, systems with fault tolerance, that consists of fault diagnosis (FD) and fault-tolerant control (FTC), are essential in practical and industrial applications, due to the great impact that the effects of the faults have on the correct performance and good conditions of physical equipment. Regarding integer-order systems, there are some surveys involving FD [1, 21, 42], while there exist some monographs concerning FTC [6, 28]. Some recent specific examples of FD and FTC for integer-order systems can be found, for instance one where a FTC approach is designed that is able to simultaneously compensate for actuator faults, model mismatch and parameter variations in aircraft systems [11]; other paper addresses the problem of

R. Martínez-Guerra et al., *Fault-tolerant Control and Diagnosis for Integer and Fractional-order Systems*, Studies in Systems, Decision and Control 328, https://doi.org/10.1007/978-3-030-62094-3_8

fault reconstruction and FTC in linear systems subject to actuator faults via learning observers, designing a reconfigurable fault-tolerant controller [17]; in another work it is proposed a FTC scheme for linear systems that uses the linear fractional transformation and LMI techniques to handle mismatched uncertainties, and adaptive techniques to compensate actuator faults [19]; in other article reduced-order functional observers are used for state estimation of linear systems with input delays, which has potential applications to fault detection [24].

Thus, given the increasing number of models with fractional dynamics, and also the importance of developing FD and FTC methods for monitoring the performance of systems, the interest in studying fractional-order systems with faults has emerged. Currently, regarding FD in fractional-order systems various approaches have been proposed; for example, a Generalized Fractional Observer Scheme that allows residual generation is used for fault detection and isolation [2]; in another work the generalized dynamic parity space method and the Luenberger diagnosis observer are used for fault detection in fractional-order thermal systems [3]; other article proposes an estimation scheme for disturbances and faults using fractional sliding mode observation [30]; another paper deals with estimation via fractional extended Kalman filter with Lévy noises [37]. Concerning fractional FTC, some other techniques have been utilized, such as an additive control used for fault-tolerance with the aid of a fractional Luenberger observer [7]; in other paper a robust FTC against uncertainties and actuator faults is proposed, where the robustness of the scheme is given via solution of LMIs [36]; in another work an Auto tuned fractional-order PID controller is used to design a FTC scheme for an Autonomous Underwater Vehicle [38].

The aim of this chapter is to present an alternative FTC methodology for a class of commensurate-order fractional nonlinear systems; the proposed scheme is hybrid, because it involves two fractional-order observers. First, FD is performed via a fractional proportional reduced-order observer plus an integral term, which has as a main advantage with respect to the approaches given in the literature that it is model-free (it does not require to know the structure of the system), and uses a property called fractional algebraic observability; also, the integral term improves the convergence of the estimation of the faults. Furthermore, for an estimation of the output tracking error a fractional high-gain observer is proposed, inspired by the one from Gauthier et al. for integer order systems [15]; the dynamics of this observer is obtained from the nominal system transformed into a fractional multi-input multi-output generalized observability canonical form (FMGOCF). Then a fractional-order dynamical controller is obtained in a natural way, whose purpose is to track desired outputs, hence eliminating the effects of the faults that enter the system. It is worth to note that this controller uses estimations of the faults obtained with the FD. Also, it is proven that the origin of the system in closed-loop with the fractional dynamical controller is Mittag-Leffler stable. Lastly, the methodology proposed is assessed with its application to two commensurate-order fractional models: the Van der Pol oscillator and a DC motor. As far as we know the papers discussed above, as well as the current literature, do not deal with the FD and FTC problems in fractional-order systems as they have been addressed in this chapter.

8.1 Fault Diagnosis

In this section is presented the condition that the faults have to satisfy for being diagnosed, and the main tool to obtain estimates of them.

In order to design a method to reconstruct the faults that appear in the system, it has to be determined if they are diagnosable, but taking into account that the system in question is of commensurate fractional order. Thus, the fractional algebraic observability property is introduced.

Definition 8.1 A variable $\eta_i \in \mathbb{R}$ is said to be fractionally algebraically observable if it is a function of the first $r_1, r_2 \in \mathbb{N}$ sequential fractional derivatives, respectively, of the known input u and the available output y, i.e.,

$$\eta_i = \phi_i(u, D^\alpha u, D^{2\alpha} u, \ldots, D^{r_1\alpha} u, y, D^\alpha y, D^{2\alpha} y, \ldots, D^{r_2\alpha} y) \qquad (8.1)$$

where $\phi_i : \mathbb{R}^{(r_1+1)m} \times \mathbb{R}^{(r_2+1)p} \to \mathbb{R}$.

Remark 8.1 If a variable fits into the above definition, it is said that it satisfies the fractional algebraic observability (FAO) condition. Both states and faults may satisfy this property. In particular, each fault that satisfies the FAO condition is said to be diagnosable.

Once being diagnosed, the faults are reconstructed by means of the fractional proportional integral reduced-order observer (FIROO) [9]:

$$D^\alpha \hat{f}_i = k_{i0}(f_i - \hat{f}_i) + \sum_{j=1}^{r'} k_{ij} I^{j\alpha}(f_i - \hat{f}_i)$$

where $\hat{f}_i$ is an estimation of the fault f_i and the terms $k_{ij} \in \mathbb{R}^+$ determine the convergence rate of the observer.

This observer considers a proportional corrective term of the fault estimation error, and fractional integral terms of the error to improve its convergence. The value r' is chosen in order to fulfill the stability conditions for the observer.

This observer has been chosen because it is model-free (it does not require to know the dynamics of the fault) and it uses the FAO condition defined to estimate the fault. In particular, in this chapter $r' = 1$ is used, that is to say

$$D^\alpha \hat{f}_i = k_{i0}(f_i - \hat{f}_i) + k_{i1} I^\alpha(f_i - \hat{f}_i). \qquad (8.2)$$

Remark 8.2 The unknown fractional dynamics of fault f_i, i.e. Ω_i, is assumed to be bounded:

$$D^\alpha f_i = \Omega_i \le \|\Omega_i\| \le N_i \qquad (8.3)$$

where $N_i \in \mathbb{R}^+$ is an upper bound for the fault dynamics.

Remark 8.3 The real fault f_i cannot be used directly since is unavailable for measurement, so its estimation $\hat{f}_i$ is used in all the existing dynamics. Due to the speed of the FIROO, the convergence time can be neglected and the estimation of the faults are used in place of the real faults.

8.2 Fault-Tolerant Control

In this section, the main methodology for constructing the FTC is presented. The nominal system is transformed into a fractional canonical form, in order to build an output error tracking fractional observer. Then, the dynamics of the controller is obtained.

The multivariable case is considered. The available outputs of the system are chosen as fractional differential primitive elements. Defining $\eta_{ij} = D^{(i-1)\alpha} y_j$, the class of nonlinear systems (7.1) can be represented by the following Fractional multi-input multi-output generalized observability canonical form (FMGOCF) [22]:

$$D^\alpha \eta_{ij} = \eta_{i+1,j}, \qquad 1 \le i \le n-1 \tag{8.4}$$
$$D^\alpha \eta_{nj} = -L_j(\eta_1, \ldots, \eta_p, u, \ldots, D^{\gamma\alpha} u, \hat{f}, \ldots, D^{\mu\alpha} \hat{f})$$
$$y_j = \eta_{1j}, \qquad 1 \le j \le p$$

where L_j is a C^1 real-valued function, $\eta_j = (\eta_{1j}, \ldots, \eta_{nj})$, $y = (y_1, \ldots, y_p)$, $u \in \mathbb{R}^m$, $\hat{f} \in \mathbb{R}^q$, and some constants $\gamma, \mu \in \mathbb{R}^+$. Actually, this FMGOCF consists in p subsystems of the form (8.4), one for each output.

The FTC proposed will consist in the tracking of the output error with respect to a given reference $y_R = (y_{R1}, \ldots, y_{Rp})$, by means of an observer-based dynamical controller built from the fractional canonical form of that error. As it can be seen, the dynamics of the canonical form (8.4) depends on the faults, so the estimations $\hat{f}_i$ obtained with the FD are used instead.

Let the output tracking error be $e_{1j} = y_j - y_{Rj}$. Given that $\eta_{ij} = D^{(i-1)\alpha} y_j$, the error variables are rewritten as:

$$e_{1j} = \eta_{1j} - y_{Rj}, \qquad 1 \le j \le p$$

Noting that

$$D^\alpha e_{1j} = D^\alpha \eta_{1j} - D^\alpha y_{Rj}$$
$$= \eta_{2j} - D^\alpha y_{Rj}$$

this change of variable defines the following fractional canonical form:

$$D^{\alpha} e_{ij} = \eta_{i+1,j} - D^{i\alpha} y_{Rj}, \qquad 1 \leq i \leq n-1 \tag{8.5}$$
$$D^{\alpha} e_{nj} = D^{\alpha} \eta_{nj} - D^{n\alpha} y_{Rj} = -L_j(\eta_1, \ldots, \eta_p, u, \ldots, D^{\gamma\alpha} u, \hat{f}, \ldots, D^{\mu\alpha} \hat{f}) - D^{n\alpha} y_{Rj}$$

Now, a linear time-invariant dynamics for the tracking error is imposed:

$$D^{\alpha} e_{nj} + \sum_{i=0}^{n-1} a_{ij} e_{ij} = 0, \qquad a_{ij} > 0 \tag{8.6}$$

From system (8.5), (8.6) is rewritten as:

$$D^{\alpha} \eta_{nj} - D^{n\alpha} y_{Rj} + \sum_{i=1}^{n} a_{ij} \left[\eta_{ij} - D^{(i-1)\alpha} y_{Rj} \right] = 0 \tag{8.7}$$

that is equivalent to

$$- L_j(\eta_1, \ldots, \eta_p, u, \ldots, D^{\gamma\alpha} u, \hat{f}, \ldots, D^{\mu\alpha} \hat{f}) - D^{n\alpha} y_{Rj} = -\sum_{i=1}^{n} a_{ij} \left[\eta_{ij} - D^{(i-1)\alpha} y_{Rj} \right] \tag{8.8}$$

A chain of integrators of the error can be obtained as follows:

$$D^{\alpha} e_{ij} = e_{i+1,j}, \qquad 1 \leq i \leq n-1$$
$$D^{\alpha} e_{nj} = -\sum_{i=1}^{n} a_{ij} e_{ij} \tag{8.9}$$

or in a vector form

$$D^{\alpha} \mathbf{e}_j = F_j \mathbf{e}_j \tag{8.10}$$

and

$$- L_j(\mathbf{e}_1 + \mathbf{y}_{R1}, \ldots, \mathbf{e}_p + \mathbf{y}_{Rp}, u, \ldots, D^{\gamma\alpha} u, \hat{f}, \ldots, D^{\mu\alpha} \hat{f}) - D^{n\alpha} y_{rj} = -\sum_{i=1}^{n} a_{ij} e_{ij} \tag{8.11}$$

where $\mathbf{e}_j = (e_{1j}, \ldots, e_{nj})$, $\mathbf{y}_{Rj} = (y_{Rj}, D^{\alpha} y_{Rj}, \ldots, D^{(n-1)\alpha} y_{Rj})$, and

$$F_j = \begin{pmatrix} 0 & 1 & \ldots & 0 \\ \vdots & \vdots & \ldots & \vdots \\ 0 & \vdots & \ldots & 1 \\ -a_{1j} & -a_{2j} & \ldots & -a_{nj} \end{pmatrix}$$

System (8.10) is stable if the following property is satisfied [29]:

$$\left| \arg \left(\lambda \left(F_j \right) \right) \right| > \alpha \frac{\pi}{2} \tag{8.12}$$

Given that dynamics (8.11) depends on the tracking errors, and only the first one is available for measurement, an observer is used to estimate the rest. Firstly, system (8.10) is rewritten as:

$$D^\alpha \mathbf{e}_j = E\mathbf{e}_j + \varphi_j \left(\mathbf{e}, \mathbf{y}_R, \mathbf{u}, \hat{\mathbf{f}} \right) \tag{8.13}$$

where $\mathbf{e} = (\mathbf{e}_1, \ldots, \mathbf{e}_p), \mathbf{y}_R = (\mathbf{y}_{R1}, \ldots, \mathbf{y}_{Rp}), \mathbf{u} = (u, \ldots, D^{\gamma\alpha}u), \hat{\mathbf{f}} = \left(\hat{f}, \ldots, D^{\mu\alpha} \hat{f} \right)$, the elements of E are given by:

$$E_{ks} = \begin{cases} 1 & \text{if } k = s - 1 \\ 0 & \text{otherwise} \end{cases}$$

and

$$\varphi_j \left(\mathbf{e}, \mathbf{y}_R, \mathbf{u}, \mathbf{f} \right) = \begin{pmatrix} 0 \\ \vdots \\ 0 \\ -L_j(\mathbf{e}_1 + \mathbf{y}_{R1}, \ldots, \mathbf{e}_p + \mathbf{y}_{Rp}, u, \ldots, D^{\gamma\alpha}u, \hat{f}, \ldots, D^{\mu\alpha} \hat{f}) - D^{n\alpha} \mathbf{y}_{Rj} \end{pmatrix}$$

The estimation $\hat{\mathbf{e}}_j$ of the tracking error vector is obtained by the following fractional high-gain observer (FHGO):

$$D^\alpha \hat{\mathbf{e}}_j = E\hat{\mathbf{e}}_j + \varphi_j \left(\hat{\mathbf{e}}, \mathbf{y}_R, \mathbf{u}, \hat{\mathbf{f}} \right) - S_\infty^{-1} C^T C(\hat{\mathbf{e}}_j - \mathbf{e}_j) \tag{8.14}$$

where S_∞ is the solution to the equation:

$$S_\infty \left(E + \frac{\theta}{2} I \right) + \left(E^T + \frac{\theta}{2} I \right) S_\infty = C^T C \tag{8.15}$$

with $\theta > 0$ and $C = \begin{pmatrix} 1 & 0 & \ldots & 0 \end{pmatrix}$. The coefficients of S_∞ are given by:

$$(S_\infty)_{ks} = \frac{a_{ks}}{\theta^{k+s-1}}$$

where (a_{ks}) is a symmetric positive definite matrix independent of θ.

Now, the dynamics of the fault-tolerant controllers is obtained from the equation of the tracking error observer (8.14). Consider the following equation:

$$- L_j(\hat{\mathbf{e}}, \mathbf{y}_R, \hat{\mathbf{u}}, D^{\gamma_l \alpha}\hat{u}_l, \hat{\mathbf{f}}) - D^{n\alpha}y_{Rj} = -\sum_{i=1}^{n} a_{ij}\hat{e}_{ij} \qquad (8.16)$$

where $D^{\gamma_l \alpha}\hat{u}_l$ is the highest order fractional derivative of the input found in the equation.

From the Implicit Function Theorem for fractional differential equations [5, 26, 40], we obtain:

$$D^{\gamma_l \alpha}\hat{u}_l = K_l\left(\hat{\mathbf{e}}, \mathbf{y}_R, D^{n\alpha}y_{Rj}, \hat{\mathbf{u}}, \hat{\mathbf{f}}\right) \qquad (8.17)$$

with solution $\hat{u}_l$, obtained numerically from a chain of integrators (see (8.19) and (8.20)); this variable represents the fault tolerant controller, with $1 \le l \le m$.

These controllers yield tracking in the original system, with fault tolerance (elimination of the effects of the faults). So, Eq. (8.14) is rewritten as:

$$D^\alpha \hat{\mathbf{e}}_j = E\hat{\mathbf{e}}_j + \varphi_j\left(\hat{\mathbf{e}}, \mathbf{y}_R, \hat{\mathbf{u}}, \hat{\mathbf{f}}\right) - S_\infty^{-1}C^T C(\hat{\mathbf{e}}_j - \mathbf{e}_j) \qquad (8.18)$$

with

$$\varphi_j\left(\hat{\mathbf{e}}, \mathbf{y}_R, \hat{\mathbf{u}}, \hat{\mathbf{f}}\right) = \begin{pmatrix} 0 \\ \vdots \\ 0 \\ -L_j(\hat{e}_1 + \mathbf{y}_{R1}, \ldots, \hat{e}_p + \mathbf{y}_{Rp}, \hat{u}, \ldots, D^{\gamma\alpha}\hat{u}, \hat{f}, \ldots, D^{\mu\alpha}\hat{f}) - D^{n\alpha}y_{Rj} \end{pmatrix}$$

If the following variables are defined:

$$\hat{u}_{il} = D^{(i-1)\alpha}\hat{u}_l \qquad i = 1, \ldots, \gamma_l \qquad (8.19)$$

then, considering the controller dynamics (8.17), the dynamical controller subsystems are written as follows:

$$D^\alpha \hat{\mathbf{u}}_l = E\hat{\mathbf{u}}_l + \kappa_l\left(\hat{\mathbf{e}}, \mathbf{y}_R, D^{n\alpha}y_{Rj}, \hat{\mathbf{u}}, \hat{\mathbf{f}}\right), \qquad 1 \le l \le m \qquad (8.20)$$

where $\hat{\mathbf{u}}_l = \left(\hat{u}_{1l}, \ldots, \hat{u}_{\gamma_l l}\right)$ and

$$\kappa_l\left(\hat{\mathbf{e}}, \mathbf{y}_R, D^{n\alpha}y_{Rj}, \hat{\mathbf{u}}, \hat{\mathbf{f}}\right) = \begin{pmatrix} 0 \\ \vdots \\ 0 \\ K_l\left(\hat{\mathbf{e}}, \mathbf{y}_R, D^{n\alpha}y_{Rj}, \hat{\mathbf{u}}, \hat{\mathbf{f}}\right) \end{pmatrix}$$

Now the following variables are defined:

$$\hat{f}_{i\bar{l}} = D^{(i-1)\alpha} \hat{f}_{\bar{l}} \quad i = 1, \ldots, \mu_{\bar{l}} \tag{8.21}$$

then, considering the FIROO dynamics (8.2), the fault estimation subsystems are written as:

$$D^{\alpha} \hat{\mathbf{f}}_{\bar{l}} = E \hat{\mathbf{f}}_{\bar{l}} + \omega_{\bar{l}}(u, y, f), \qquad 1 \le \bar{l} \le q \tag{8.22}$$

where $\hat{\mathbf{f}}_{\bar{l}} = \left(\hat{f}_{1\bar{l}}, \ldots, \hat{f}_{\mu_{\bar{l}}\bar{l}} \right)$ and

$$\omega_{\bar{l}}(u, y, f) = \begin{pmatrix} 0 \\ \vdots \\ 0 \\ D^{(\mu_{\bar{l}}-1)\alpha} k_{\bar{l}0}(f_{\bar{l}} - \hat{f}_{\bar{l}}) + D^{(\mu_{\bar{l}}-1)\alpha} k_{\bar{l}1}(I^{\alpha} f_{\bar{l}} - I^{\alpha} \hat{f}_{\bar{l}}) \end{pmatrix}.$$

Remark 8.4 Similar to the integer-order case, the overall FD and FTC system consists on a FIROO and a FHGO, which represents a hybrid observation system. From this closed-loop system, a separation principle can be determined.

8.3 Stability Analysis of the Closed-Loop System

Hence, the closed-loop dynamics is given by:

$$D^{\alpha} \hat{\mathbf{e}}_j = E \hat{\mathbf{e}}_j + \varphi_j \left(\hat{\mathbf{e}}, \mathbf{y}_R, D^{n\alpha} y_{Rj}, \hat{\mathbf{u}}, \hat{\mathbf{f}} \right) - S_{\infty}^{-1} C^T C (\hat{\mathbf{e}}_j - \mathbf{e}_j) \tag{8.23}$$

$$D^{\alpha} \hat{\mathbf{u}}_l = E \hat{\mathbf{u}}_l + \kappa_l \left(\hat{\mathbf{e}}, \mathbf{y}_R, D^{n\alpha} y_{Rj}, \hat{\mathbf{u}}, \hat{\mathbf{f}} \right)$$

$$D^{\alpha} \hat{\mathbf{f}}_{\bar{l}} = E \hat{\mathbf{f}}_{\bar{l}} + \omega_{\bar{l}}(u, y, f)$$

for $1 \le j \le p$, $1 \le l \le m$ and $1 \le \bar{l} \le q$. Writing explicitly these equations, the following chain of integrators is obtained:

$$D^{\alpha} \hat{e}_{ij} = \hat{e}_{i+1,j} - \psi_i(\theta_j)(\hat{e}_j - e_j) \quad 1 \le i \le n-1$$

$$D^{\alpha} \hat{e}_{nj} = -L_j(\hat{\mathbf{e}}_1 + \mathbf{y}_{R1}, \ldots, \hat{\mathbf{e}}_p + \mathbf{y}_{Rp}, \hat{u}, \ldots, D^{\gamma\alpha} \hat{u}, \hat{f}, \ldots, D^{\mu\alpha} \hat{f}) - D^{n\alpha} y_{Rj} - \theta_j^n$$

$$\qquad 1 \le j \le p$$

$$D^{\alpha} \hat{u}_{il} = \hat{u}_{i+1,l} \quad 1 \le i \le \gamma_l - 1$$

$$D^{\alpha} \hat{u}_{\gamma_l,l} = K_l \left(\hat{\mathbf{e}}, \mathbf{y}_R, D^{n\alpha} y_{Rj}, \hat{\mathbf{u}}, \hat{\mathbf{f}} \right) \qquad 1 \le l \le m$$

$$D^{\alpha} \hat{f}_{i,\bar{l}} = \hat{f}_{i+1,\bar{l}} \quad 1 \le i \le \mu_{\bar{l}} - 1$$

$$D^{\alpha} \hat{f}_{\mu_{\bar{l}},\bar{l}} = D^{(\mu_{\bar{l}}-1)\alpha} k_{\bar{l}0}(f_{\bar{l}} - \hat{f}_{\bar{l}}) + D^{(\mu_{\bar{l}}-1)\alpha} k_{\bar{l}1}(I^{\alpha} f_{\bar{l}} - I^{\alpha} \hat{f}_{\bar{l}}) \qquad 1 \le \bar{l} \le q$$

where $\psi_i(\theta_j)$ is a function obtained from S_{∞}^{-1}.

In this chain of integrators, the dynamics of the controllers and the fault estimations can be appreciated. As it can be seen, the variables obtained from these dynamics take part explicitly in the tracking error dynamics, leading to the solution of the tracking problem.

Moreover, define the observation error as $\varepsilon_j = \hat{\mathbf{e}}_j - \mathbf{e}_j$, and the following dynamics is obtained from Eqs. (8.14) and (8.18):

$$D^\alpha \varepsilon_j = \left(E - S_\infty^{-1} C^T C\right) \varepsilon_j + \Phi_j(\varepsilon, \hat{\mathbf{e}}) \tag{8.24}$$

where

$$\Phi_j(\varepsilon, \hat{\mathbf{e}}) = \varphi_j\left(\hat{\mathbf{e}}, \mathbf{y}_R, \hat{\mathbf{u}}, \hat{\mathbf{f}}\right) - \varphi_j\left(\hat{\mathbf{e}} - \varepsilon, \mathbf{y}_R, \hat{\mathbf{u}}, \hat{\mathbf{f}}\right)$$

Now the main result of this chapter is stated.

Theorem 8.1 *Let system (7.1) be described in the FMGOCF (8.4) composed of p subsystems. Let the observation dynamics corresponding to subsystem j be composed of $\hat{\mathbf{e}}_j$ and ε_j, for $1 \leq j \leq p$. Let $f_{\bar{l}}$ be diagnosable and estimated by means of the dynamics of $\hat{f}_{\bar{l}}$, for $1 \leq \bar{l} \leq q$. Let $\hat{u}_l$ be the solution to*

$$-L_j(\hat{\mathbf{e}}, \mathbf{y}_R, \hat{\mathbf{u}}, D^{\gamma_l \alpha} \hat{u}_l, \hat{\mathbf{f}}) - y_{Rj}^{(n\alpha)} = -\sum_{i=1}^{n} a_{ij} \hat{e}_{ij}$$

for $1 \leq l \leq m$. Then, the origin of the closed-loop system (8.23) is Mittag-Leffler stable.

Proof Consider the following Lyapunov function candidate:

$$V\left(\hat{\mathbf{e}}_j, \varepsilon_j, \tilde{f}_{\bar{l}}\right) = \hat{\mathbf{e}}_j^T P \hat{\mathbf{e}}_j + \varepsilon_j^T S_\infty \varepsilon_j + \tilde{f}_{\bar{l}}^2 \tag{8.25}$$

where $\tilde{f}_{\bar{l}} = f_{\bar{l}} - \hat{f}_{\bar{l}}$ is the fault estimation error. Define $V_1\left(\hat{\mathbf{e}}_j\right) = \hat{\mathbf{e}}_j^T P \hat{\mathbf{e}}_j$, $V_2\left(\varepsilon_j\right) = \varepsilon_j^T S_\infty \varepsilon_j$ and $V_3(\tilde{f}_{\bar{l}}) = \tilde{f}_{\bar{l}}^2$. P, S_∞ are constant, square, symmetric, positive definite matrices and solutions of $F^T P + PF = -I$ and (8.15) respectively. Let $\|x\|_P = \sqrt{x^T P x}$ and $\|x\|_{S_\infty} = \sqrt{x^T S_\infty x}$. Note that $V\left(\hat{\mathbf{e}}_j, \varepsilon_j, \tilde{f}_{\bar{l}}\right)$ satisfies the first inequality of Theorem 7.1 since:

$$\alpha_{11}\|\hat{\mathbf{e}}_j\| \leq V_1\left(\hat{\mathbf{e}}_j\right) \leq \alpha_{21}\|\hat{\mathbf{e}}_j\| \tag{8.26}$$
$$\alpha_{12}\|\varepsilon_j\| \leq V_2\left(\varepsilon_j\right) \leq \alpha_{22}\|\varepsilon_j\|$$
$$\alpha_{13}\|\tilde{f}_{\bar{l}}\| \leq V_3(\tilde{f}_{\bar{l}}) \leq \alpha_{23}\|\tilde{f}_{\bar{l}}\|$$

with $\alpha_{11} = \lambda_{\min}(P)$, $\alpha_{12} = \lambda_{\min}(S_\infty)$, $\alpha_{13} = 1$, $\alpha_{21} = \frac{1}{2}(\lambda_{\min}(P) + \lambda_{\max}(P))$, $\alpha_{22} = \frac{1}{2}(\lambda_{\min}(S_\infty) + \lambda_{\max}(S_\infty))$, $\alpha_{23} = \sup(\|\tilde{f}_{\bar{l}}\|)$, and $a = b = 1$.

By the linearity property of the Caputo Derivative and using Lemma 7.1, it follows that

$$D^\alpha V = D^\alpha V_1 + D^\alpha V_2 + D^\alpha V_3 \leq 2\hat{\mathbf{e}}_j^T P D^\alpha \hat{\mathbf{e}}_j + 2\varepsilon_j^T S_\infty D^\alpha \varepsilon_j + 2\tilde{f}_{\bar{l}} D^\alpha \tilde{f}_{\bar{l}} \quad (8.27)$$

On the other hand note that:

$$
\begin{aligned}
D^\alpha V_1 \left(\hat{\mathbf{e}}_j\right) &\leq 2\hat{\mathbf{e}}_j^T P D^\alpha \hat{\mathbf{e}}_j = 2\hat{\mathbf{e}}_j^T P \left[F\hat{\mathbf{e}}_j - S_\infty^{-1} C^T C \varepsilon_j\right] \\
&= \hat{\mathbf{e}}_j^T \left(F^T P + P F\right) \hat{\mathbf{e}}_j - 2\hat{\mathbf{e}}_j^T P S_\infty^{-1} C^T C \varepsilon_j \\
&= -\|\hat{\mathbf{e}}_j\|^2 - 2\hat{\mathbf{e}}_j^T P S_\infty^{-1} C^T C S_\infty^{-1} S_\infty \varepsilon_j \\
&\leq -\|\hat{\mathbf{e}}_j\|^2 + \bar{K}\|\hat{\mathbf{e}}_j\|_{P^*}\|\varepsilon_j\|_{S_\infty^*}\rho(\theta) \\
&\leq -\left(1 - \bar{K} d_1 d_2 \|\varepsilon_j\|\rho(\theta)\right)\|\hat{\mathbf{e}}_j\|
\end{aligned}
$$

where $\bar{K} > 0$, $P^* = PP^T$, $S_\infty^* = S_\infty S_\infty^T$, $d_1 = \sqrt{\lambda_{\max}\left(P^2\right)}$, $d_2 = \sqrt{\lambda_{\max}\left(S_\infty^2\right)}$ and $\rho(\theta) = \|S_\infty^{-1} C^T C S_\infty^{-1}\|$.

Then, we obtain

$$D^\alpha V_1 \left(\hat{\mathbf{e}}_j\right) \leq -\delta_{31}\|\hat{\mathbf{e}}_j\| \quad (8.28)$$

with $\delta_{31} = 1 - \bar{K} d_1 d_2 \|\varepsilon_j\|\rho(\theta)$.

In a similar way:

$$
\begin{aligned}
D^\alpha V_2 \left(\varepsilon_j\right) &\leq 2\varepsilon_j^T S_\infty D^\alpha \varepsilon_j = 2\varepsilon_j^T S_\infty \left[E_\theta \varepsilon_j + \Phi_j\left(\varepsilon, \hat{\mathbf{e}}\right)\right] \\
&= \varepsilon_j^T \left[E^T S_\infty + S_\infty E - C^T C\right] \varepsilon_j - \varepsilon_j^T C^T C \varepsilon_j + 2\varepsilon_j^T S_\infty \Phi_j\left(\varepsilon, \hat{\mathbf{e}}\right) \\
&\leq -\theta\|\varepsilon_j\|_{S_\infty}^2 + 2\varepsilon_j^T S_\infty \Phi_j\left(\varepsilon, \hat{\mathbf{e}}\right) \quad \text{(since } \|\Phi_j\left(\varepsilon, \hat{\mathbf{e}}\right)\|_{S_\infty}^2 \leq \bar{\lambda}\|\varepsilon_j\|_{S_\infty}^2, \ \bar{\lambda} = \gamma^2\text{)} \\
&\leq -\theta\lambda_{\min}\left(S_\infty\right)\|\varepsilon_j\|^2 + 2\gamma\|\varepsilon_j\|_{S_\infty}^2 \\
&\leq -\left(\theta\lambda_{\min}\left(S_\infty\right) - 2\gamma\lambda_{\max}\left(S_\infty\right)\right)\|\varepsilon_j\|
\end{aligned}
$$

Then, we obtain

$$D^\alpha V_2 \left(\varepsilon_j\right) \leq -\delta_{32}\|\varepsilon_j\| \quad (8.29)$$

with $\delta_{32} = \theta\lambda_{\min}\left(S_\infty\right) - 2\gamma\lambda_{\max}\left(S_\infty\right)$.

Finally:

$$
\begin{aligned}
D^\alpha V_3(\tilde{f}_{\bar{l}}) &\leq 2\tilde{f}_{\bar{l}} D^\alpha \tilde{f}_{\bar{l}} = 2\tilde{f}_{\bar{l}}\left(\Omega_{\bar{l}} - k_{\bar{l}0}\tilde{f}_{\bar{l}} - k_{\bar{l}1} I^\alpha \tilde{f}_{\bar{l}}\right) \\
&\leq 2\tilde{f}_{\bar{l}}\Omega_{\bar{l}} - 2k_{\bar{l}1}\tilde{f}_{\bar{l}} I^\alpha \tilde{f}_{\bar{l}} \\
&\leq 2N_{\bar{l}}\|\tilde{f}_{\bar{l}}\| - 2k_{\bar{l}1}\|\tilde{f}_{\bar{l}}\||I^\alpha \tilde{f}_{\bar{l}}| \\
&\leq -\left(2k_{\bar{l}1}|I^\alpha \tilde{f}_{\bar{l}}| - 2N_{\bar{l}}\right)\|\tilde{f}_{\bar{l}}\|
\end{aligned}
$$

Then, we obtain

$$D^\alpha V_3(\tilde{f}_{\bar{l}}) \leq -\delta_{33}\|\tilde{f}_{\bar{l}}\| \quad (8.30)$$

with $\delta_{33} = 2k_{\tilde{i}1}|I^{\alpha}\tilde{f}_{\tilde{i}}| - 2N_{\tilde{i}}$.

Therefore, if δ_{31}, δ_{32}, $\delta_{33} > 0$, from Theorem 7.1 and Eqs. (8.26)–(8.30), it is concluded that the origin of system (8.23) is Mittag-Leffler stable.   □

8.4   Application

8.4.1   Fractional van der Pol Oscillator

Consider the modified version of the fractional-order Van der Pol oscillator [29]:

$$D^{\alpha}x_1 = x_2$$
$$D^{\alpha}x_2 = -x_1 - \varepsilon(x_1^2 - 1)x_2$$

Adding a control input and a fault and selecting the first state as a measurable output to be controlled, the system to work with is:

$$D^{\alpha}x_1 = x_2 + u \qquad (8.31)$$
$$D^{\alpha}x_2 = -x_1 - \varepsilon(x_1^2 - 1)x_2 + f$$
$$y = x_1$$

In this case, the control aim is tracking of a desired trajectory by the output (in this case the first state), thus making the oscillator to present a desired chaotic behavior in the phase plane, even in the presence of faults.

8.4.1.1   Fault Diagnosis

First, the FIROO for reconstructing the fault is designed, for which it must be determined if f is diagnosable, i.e. if it satisfies the FAO condition. From (8.31) the following polynomial is obtained:

$$f = D^{\alpha}x_2 + y + \varepsilon(y^2 - 1)x_2 \qquad (8.32)$$

It can be observed that state x_2 satisfies also the FAO condition. From (8.31):

$$x_2 = D^{\alpha}y - u \qquad (8.33)$$

thus, state x_2 is fractionally algebraically observable and can be reconstructed. Hence, the fault is fractionally diagnosable. First, a FIROO for estimating x_2 is designed:

$$D^{\alpha}\hat{x}_2 = k_{10}(D^{\alpha}y - u - \hat{x}_2) + k_{11}I^{\alpha}(D^{\alpha}y - u - \hat{x}_2)$$

Defining an auxiliary variable γ_1 as:

$$\gamma_1 = \hat{x}_2 - k_{10}y$$

the FIROO to obtain the estimation of the state is:

$$D^\alpha \gamma_1 = -k_{10}(u + \gamma_1 + k_{10}y) + k_{11}y - k_{11}I^\alpha(u + \gamma_1 + k_{10}y) \tag{8.34}$$

$$\hat{x}_2 = \gamma_1 + k_{10}y \tag{8.35}$$

Now, the FIROO for estimating the fault is designed:

$$D^\alpha \hat{f} = k_{20}(D^\alpha \hat{x}_2 + y + \varepsilon(y^2 - 1)\hat{x}_2 - \hat{f}) + k_{21}I^\alpha(D^\alpha \hat{x}_2 + y + \varepsilon(y^2 - 1)\hat{x}_2 - \hat{f})$$

Defining an auxiliary variable γ_2 as

$$\gamma_2 = \hat{f} - k_{20}\hat{x}_2$$

the FIROO to obtain the estimation of the fault is:

$$D^\alpha \gamma_2 = -k_{20}(-y - \varepsilon(y^2 - 1)\hat{x}_2 + \gamma_2 + k_{20}\hat{x}_2) + k_{21}\hat{x}_2 \tag{8.36}$$
$$-k_{21}I^\alpha(-y - \varepsilon(y^2 - 1)\hat{x}_2 + \gamma_2 + k_{20}\hat{x}_2)$$

$$\hat{f} = \gamma_2 + k_{20}\hat{x}_2. \tag{8.37}$$

8.4.1.2 Fault-Tolerant Control

Now, a fractional fault-tolerant controller is designed from the FMGOCF, obtaining the following:

$$D^\alpha y = x_2 + u \tag{8.38}$$
$$D^{2\alpha} y = D^\alpha x_2 + D^\alpha u$$
$$= -y - \varepsilon(y^2 - 1)x_2 + D^\alpha u + f$$

So the tracking error is described in a canonical form:

$$e_1 = y - y_R \tag{8.39}$$
$$D^\alpha e_1 = D^\alpha y - D^\alpha y_R = e_2$$
$$D^\alpha e_2 = D^{2\alpha} e_1 = D^{2\alpha} y - D^{2\alpha} y_R$$

and a FHGO can be obtained:

$$D^\alpha \hat{e}_1 = \hat{e}_2 - 2\theta(\hat{e}_1 - e_1) \tag{8.40}$$
$$D^\alpha \hat{e}_2 = D^{2\alpha} y - D^{2\alpha} y_R - \theta^2(\hat{e}_1 - e_1)$$
$$= -y - \varepsilon(y^2 - 1)x_2 + D^\alpha u + \hat{f} - D^{2\alpha} y_R - \theta^2(\hat{e}_1 - e_1) = -\sum_{i=1}^{2} a_i \hat{e}_i$$

Finally, from Eq. (8.40), the dynamics of the fractional fault-tolerant controller is obtained:

$$D^\alpha \hat{u} = -a_1 \hat{e}_1 - a_2 \hat{e}_2 + y + \varepsilon(y^2 - 1)\hat{x}_2 - \hat{f} + D^{2\alpha} y_R. \tag{8.41}$$

8.4.1.3 Simulation Results

Simulations were performed with the Ninteger Toolbox from Matlab-Simulink®, over 60 s in the model of the system. The value $\alpha = 0.9$ is selected. The reference is set as $y_R(t) = 2\sin(t)$. The fault is set to be $f(t) = \cos(t)$ beginning at 20 s. The value of the parameter of the system is $\varepsilon = 0.1$. The design parameters (gains) are chosen as $\theta = 20$, $a_1 = 400$, $a_2 = 40$, $k_{10} = k_{20} = 30$ and $k_{11} = k_{21} = 1$.

Figure 8.1 shows the FD results with the FIROO. It can be seen that the estimated fault follows the signal of the real fault in a very short time. The performance index of the FIROO was evaluated using the following cost functional:

$$J_t = \frac{1}{t + \varepsilon} \int_0^t \left\| \tilde{f} \right\|^2 dt$$

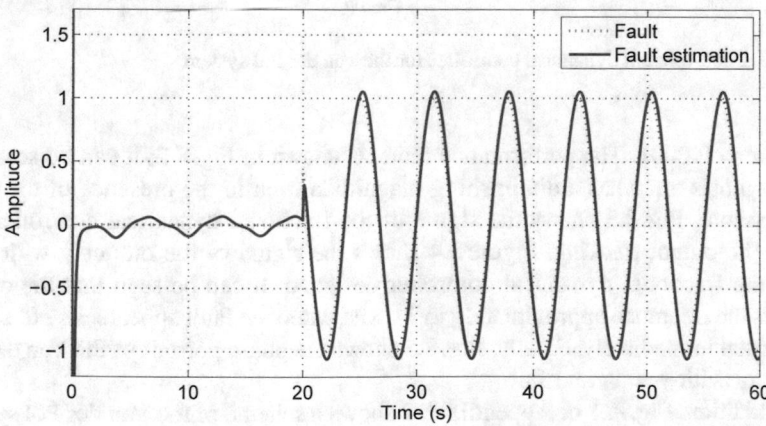

Fig. 8.1 Fault diagnosis for the Van der Pol system

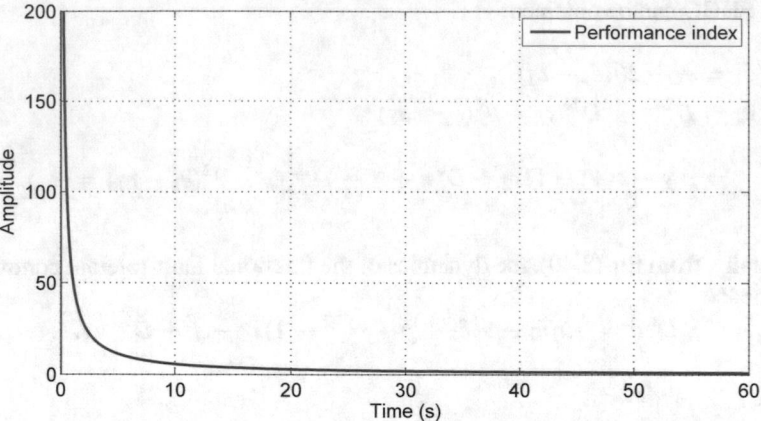

Fig. 8.2 Performance index of the fault diagnosis for the Van der Pol system

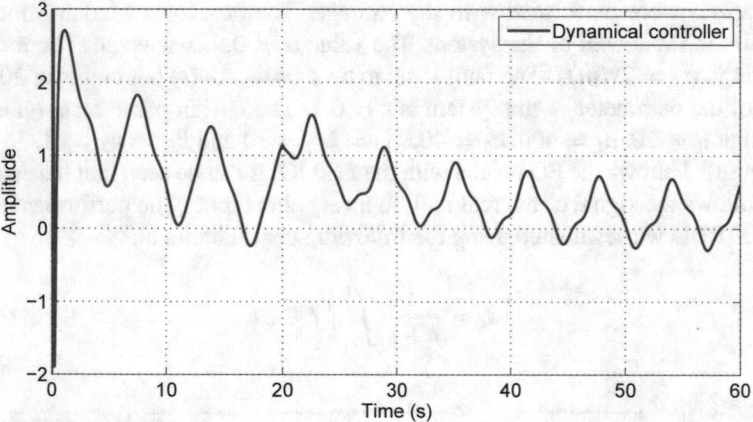

Fig. 8.3 Fault-tolerant dynamical controller for the Van der Pol system

where $\varepsilon = 0.0001$. The performance index is shown in Fig. 8.2; it can be seen that the diagnosis error has a diminishing magnitude even in the presence of the fault. Furthermore, Fig. 8.3 shows the signal of the fractional dynamical controller that yields the output tracking. Figure 8.4 shows the signal of the output y with FTC using the fractional dynamical controller designed. It can be seen that the output follows the reference approximately in 1 s, and when the fault appears, its effects are eliminated immediately. Finally, Fig. 8.5 shows the phase portrait of the Van der Pol oscillator with $y = x_1$ and $\hat{x}_2$.

In addition, Fig. A.1 of Appendix A.2 shows a scheme of the Van der Pol system in closed-loop with the fractional fault diagnosis observer and the fractional fault-tolerant dynamical controller (Matlab® Simulink).

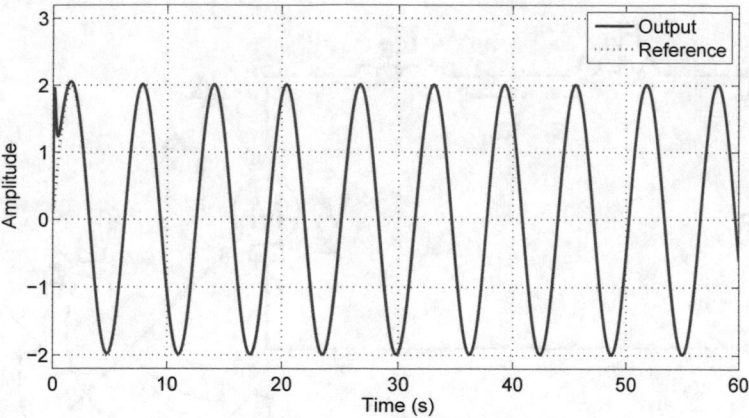

Fig. 8.4 Output tracking for the Van der Pol system

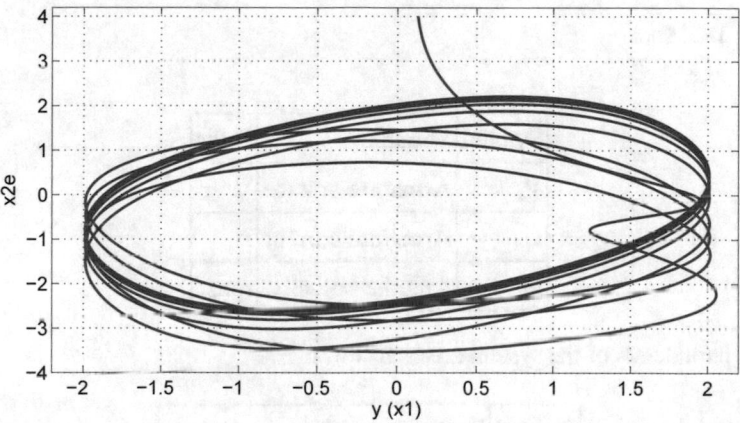

Fig. 8.5 Phase portrait of the Van der Pol system

8.4.2 Fractional Model of a DC Motor

Here is presented a fractional model of a DC motor (Fig. 8.6):

$$D^{\alpha}x(t) = \omega(t) \tag{8.42}$$

$$D^{\alpha}\omega(t) = \frac{1}{J}[c\phi i_a(t) - T_L]$$

$$D^{\alpha}i_a(t) = \frac{1}{L_a}[V_a - R_a i_a(t) - c\phi\omega(t)]$$

The variables of the system are given by

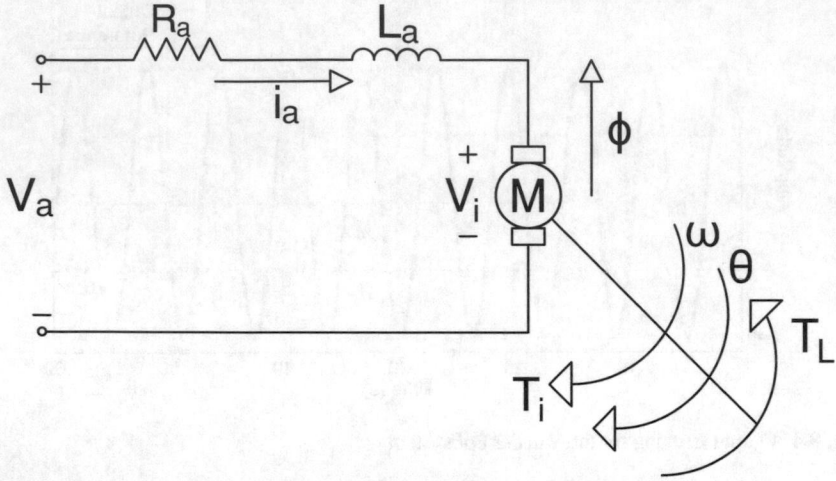

Fig. 8.6 DC Motor

Symbol	Variable	Units
V_a	Armature voltage	V
i_a	Armature current	A
ω	Angular velocity	rpm

The parameters of the system are as follows:

Symbol	Parameter	Units
R_a	Armature resistance	Ω
L_a	Armature inductance	H
ϕ	Magnetic flux	Vs
J	Total moment of inertia	kgm^2
T_L	Load torque	Nm

Besides

$$c\phi\omega(t) = V_i \qquad c\phi i_a(t) = T_i$$

where $c\phi$ is a motor constant, V_i is the induced voltage and T_i is the electromagnetic torque. Futhermore, consider an state $x(t)$ such that $I^\alpha \omega = x$ (augmented system of [8]).

Remark 8.5 If α is set as 1 in this integral, x would be the angular position, the integral of ω. Since $\alpha \neq 1$, the variable x does not represent the angular position, but only the fractional-order integral of ω.

Then, the state-space variables are chosen as $x_1 = x$, $x_2 = \omega$, $x_3 = i_a$, $u = V_a$ and $y = \omega$. Also, consider an additive fault f coupled to the input. So, the model to be used is:

$$D^\alpha x_1 = x_2 \tag{8.43}$$
$$D^\alpha x_2 = \frac{1}{J}[c\phi x_3 - T_L]$$
$$D^\alpha x_3 = \frac{1}{L_a}[-c\phi x_2 - R_a x_3 + u + f]$$
$$y = x_2$$

It can be seen that the control aim is to maintain the motor speed in the nominal value, even in the presence of faults, which in this case are variations in the input voltage.

8.4.2.1 Fault Diagnosis

Now, the FIROO for reconstructing the fault is designed, but first it must be determined if f is diagnosable, i.e. if it satisfies the FAO condition. From Eq. (8.43) the following polynomial is obtained:

$$f = c\phi y + R_a x_3 + L_a D^\alpha x_3 - u \tag{8.44}$$

It can be observed that state x_3 satisfies also the FAO condition. From (8.43):

$$x_3 = \frac{1}{c\phi}\left[J D^\alpha y + T_L\right] \tag{8.45}$$

thus, state x_3 is fractionally algebraically observable and can be reconstructed. Hence, the fault is fractionally diagnosable. Now, the FIROO is designed:

$$D^\alpha \hat{f} = k_{10}(f - \hat{f}) + k_{11} I^\alpha (f - \hat{f}) \tag{8.46}$$
$$= k_{10}(c\phi y + R_a \hat{x}_3 + L_a D^\alpha \hat{x}_3 - u - \hat{f})$$
$$+ k_{11} I^\alpha (c\phi y + R_a \hat{x}_3 + L_a D^\alpha \hat{x}_3 - u - \hat{f})$$

In order to deal with the fractional derivative of $\hat{x}_3$ in the proportional term, the auxiliary variable γ_1 is proposed and the estimated fault is defined as:

$$\hat{f} = \gamma_1 + k_{10} L_a \hat{x}_3 \tag{8.47}$$

and

$$D^\alpha \gamma_1 = D^\alpha \hat{f} - k_{10} L_a D^\alpha \hat{x}_3 \tag{8.48}$$
$$= k_{10}(c\phi y + R_a \hat{x}_3 + L_a D^\alpha \hat{x}_3 - u - \hat{f})$$
$$+ k_{11} I^\alpha (c\phi y + R_a \hat{x}_3 + L_a D^\alpha \hat{x}_3 - u - \hat{f}) - k_{10} L_a D^\alpha \hat{x}_3$$

So, the FIROO to estimate the fault is:

$$D^\alpha \gamma_1 = k_{10}(c\phi y + R_a \hat{x}_3 - u - \gamma_1 - k_{10} L_a \hat{x}_3) \tag{8.49}$$
$$+ k_{11} I^\alpha (c\phi y + R_a \hat{x}_3 + L_a D^\alpha \hat{x}_3 - u - \gamma_1 - k_{10} L_a \hat{x}_3)$$
$$\hat{f} = \gamma_1 + k_{10} L_a \hat{x}_3$$

where $\hat{x}_3$ is given by:

$$\hat{x}_3 = \frac{1}{c\phi} \left[J D^\alpha y + T_L \right] \tag{8.50}$$

In order to estimate the α-order derivative of y, the variable $\xi = D^\alpha y$ is defined and the following FIROO is proposed:

$$D^\alpha \hat{\xi} = k_{\xi 0}(\xi - \hat{\xi}) + k_{\xi 1} I^\alpha (\xi - \hat{\xi}) \tag{8.51}$$
$$= k_{\xi 0}(D^\alpha y - \hat{\xi}) + k_{\xi 1} I^\alpha (D^\alpha y - \hat{\xi})$$

Now the auxiliary variable γ_ξ is introduced and the estimated variable $\hat{\xi}$ is defined as:

$$\hat{\xi} = \gamma_\xi + k_{\xi 0} y \tag{8.52}$$

Then

$$D^\alpha \gamma_\xi = D^\alpha \hat{\xi} - k_{\xi 0} D^\alpha y \tag{8.53}$$
$$= -k_{\xi 0} \hat{\xi} + k_{\xi 1} I^\alpha (D^\alpha y - \hat{\xi})$$
$$= k_{\xi 0} \left(-\gamma_\xi - k_{\xi 0} y \right) + k_{\xi 1} I^\alpha (D^\alpha y - \gamma_\xi - k_{\xi 0} y)$$

Hence, the estimation of $\hat{x}_3$ is obtained with:

$$\hat{x}_3 = \frac{1}{c\phi} \left[J \hat{\xi} + T_L \right]. \tag{8.54}$$

8.4.2.2 Fault-Tolerant Control

Now, a fractional fault-tolerant controller is designed from the FMGOCF, obtaining the following:

$$D^{\alpha} y = \frac{1}{J} [c\phi x_3 - T_L] \tag{8.55}$$

$$D^{2\alpha} y = \frac{c\phi}{J} D^{\alpha} x_3 = \frac{c\phi}{L_a J} [-c\phi x_2 - R_a x_3 + u + f]$$

$$D^{3\alpha} y = \frac{c\phi}{L_a J} \left[-c\phi D^{\alpha} x_2 - R_a D^{\alpha} x_3 + D^{\alpha} u + D^{\alpha} f \right]$$

$$= \frac{c\phi}{L_a J} \left[R_a c\phi y - c\phi D^{\alpha} y + R_a^2 x_3 - R_a u - R_a f + D^{\alpha} u + D^{\alpha} f \right]$$

So the tracking error is described in a canonical form:

$$e_1 = y - y_R \tag{8.56}$$
$$D^{\alpha} e_1 = D^{\alpha} y - D^{\alpha} y_R = e_2$$
$$D^{\alpha} e_2 = D^{2\alpha} e_1 = D^{2\alpha} y - D^{2\alpha} y_R$$
$$D^{\alpha} e_3 = D^{3\alpha} e_1 = D^{3\alpha} y - D^{3\alpha} y_R$$

and a FHGO can be obtained:

$$D^{\alpha} \hat{e}_1 = \hat{e}_2 - 3\theta (\hat{e}_1 - e_1) \tag{8.57}$$
$$D^{\alpha} \hat{e}_2 = \hat{e}_3 - 3\theta^2 (\hat{e}_1 - e_1)$$
$$D^{\alpha} \hat{e}_3 = \frac{c\phi}{L_a J} \left[R_a c\phi y - c\phi D^{\alpha} y + R_a^2 \hat{x}_3 - R_a \hat{u} - R_a \hat{f} + D^{\alpha} \hat{u} + D^{\alpha} \hat{f} \right] - D^{3\alpha} y_R$$
$$-\theta^3 (\hat{e}_1 - e_1) = - \sum_{i=1}^{3} a_i \hat{e}_i$$

From Eq. (8.57), the dynamics of the fractional fault-tolerant controller is obtained:

$$D^{\alpha} \hat{u} = \frac{L_a J}{c\phi} \left(-a_1 \hat{e}_1 - a_2 \hat{e}_2 - a_3 \hat{e}_3 + D^{3\alpha} y_R \right) - R_a c\phi y + c\phi y^{(\alpha)} - R_a^2 \hat{x}_3 + R_a \hat{u} + R_a \hat{f} - D^{\alpha} \hat{f} \tag{8.58}$$

This controller eliminates the effects of the fault in the system; however, it can be seen that an estimation of the α-order derivative of y is needed. Thus, the dynamics of the fault-tolerant controller is the following:

$$D^{\alpha} \hat{u} = \frac{L_a J}{c\phi} \left(-a_1 \hat{e}_1 - a_2 \hat{e}_2 - a_3 \hat{e}_3 + D^{3\alpha} y_R \right) - R_a c\phi y + c\phi \hat{\xi} - R_a^2 \hat{x}_3 + R_a \hat{u} + R_a \hat{f} - D^{\alpha} \hat{f}. \tag{8.59}$$

Remark 8.6 Note that the dynamics of $D^{\alpha} \hat{f}$ is obtained from the dynamics of the FIROO used for FD.

8.4.2.3   Simulation Results

Simulations were performed over 20 s in the model of the system, with the following values of the parameters:

R_a	$2.13\,\Omega$
L_a	$0.00484\,\mathrm{H}$
$c\phi$	$0.0683\,\mathrm{Vs}$
J	$0.0001148\,\mathrm{kgm}^2$
T_L	$0.0608\,\mathrm{Nm}$
V_a	$12\,\mathrm{V}$

The value $\alpha = 0.9$ is selected. The reference is set as $y_R = 177$ rpm. The fault is set to be $f = 0.1V_a$ beginning at 10 s. The dimensionless design parameters (gains) are chosen as $\theta = 2000, a_1 = 8000, a_2 = 1200, a_3 = 60, k_{10} = k_{\xi 0} = 20$ and $k_{11} = k_{\xi 1} = 100$.

Figure 8.7 shows the FD results with the FIROO. It can be seen that the estimated fault follows the signal of the real fault in a short time. The performance index of the FIROO was evaluated using the same cost function

$$J_t = \frac{1}{t + \varepsilon} \int_0^t \left\| \tilde{f} \right\|^2 dt$$

with $\varepsilon = 0.0001$. The performance index is shown in Fig. 8.8; it can be seen that the diagnosis error has a small magnitude even in the presence of the fault. Furthermore, Fig. 8.9 shows the signal of the fractional dynamical controller that yields the output tracking. Finally Fig. 8.10 shows the signal of the output y with FTC using the fractional dynamical controller. It can be seen that the system follows the reference in 3 s, and when the fault appears, its effects are eliminated in approximately 2 s.

In addition, Fig. A.2 of Appendix A.2 shows a scheme of the motor system in closed-loop with the fractional fault diagnosis observer and the fractional fault-tolerant dynamical controller (Matlab® Simulink).

8.4.3   Comparison of the Fractional-Order DC Motor with the Integer-Order Case

In this section, a comparison is made between the results obtained in the former section (fractional-order FTC) with the ones obtained applying the methodology to the DC motor but with integer-order dynamics, for which the scheme proposed in Chap. 4 is used.

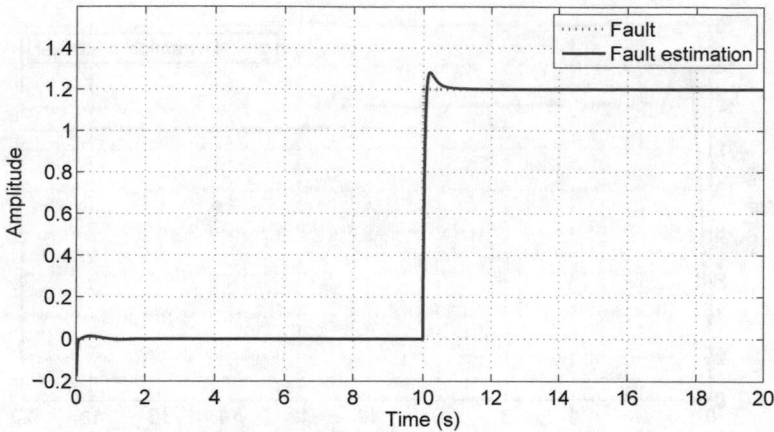

Fig. 8.7 Fault diagnosis for the DC motor

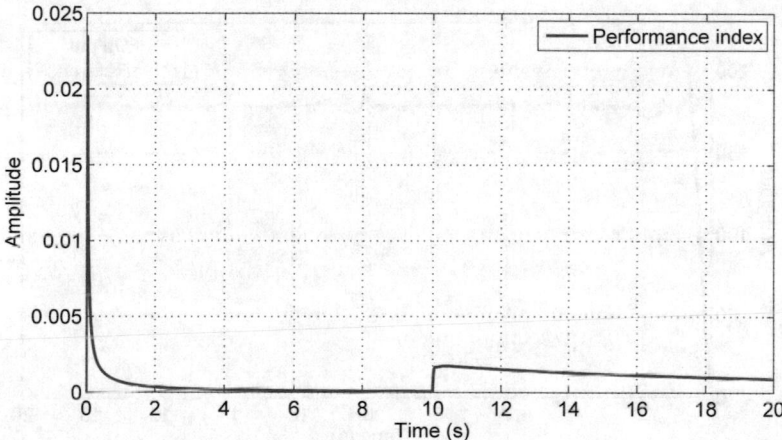

Fig. 8.8 Performance index of the fault diagnosis for the DC motor

Consider the integer order model of the DC motor:

$$\dot{\theta} = \omega \qquad (8.60)$$
$$\dot{\omega} = \frac{1}{J}[c\phi i_a - T_L]$$
$$\dot{i}_a = \frac{1}{L_a}[V_a - R_a i_a - c\phi\omega].$$

The system variables are:

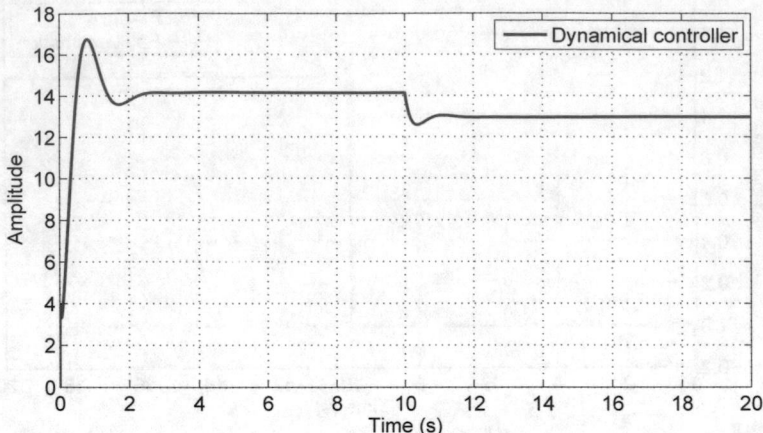

Fig. 8.9 Fault-tolerant dynamical controller for the DC motor

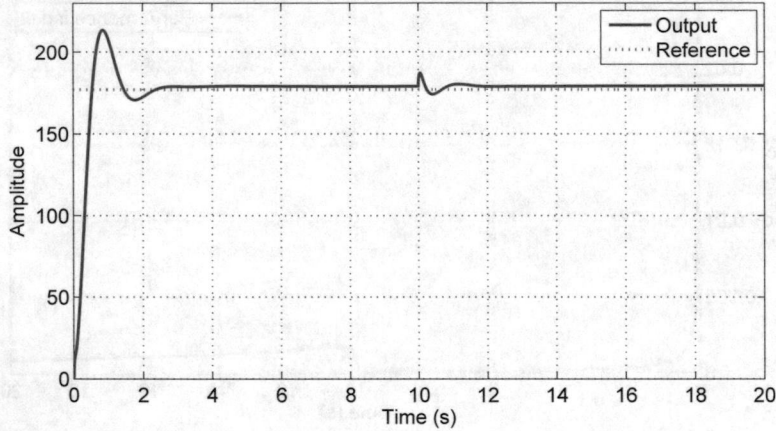

Fig. 8.10 Output tracking for the DC motor

Symbol	Variable	Units
V_a	Armature voltage	V
i_a	Armature current	A
θ	Angular position	rad
ω	Angular velocity	rpm

The system parameters are:

Symbol	Parameter	Units
R_a	Armature resistance	Ω
L_a	Armature inductance	H
ϕ	Magnetic flux	Vs
J	Total moment of inertia	kgm^2
T_L	Load torque	Nm

The state variables are chosen as $x_1 = \theta$, $x_2 = \omega$, $x_3 = i_a$, $u = V_a$ and $y = \omega$. Besides, consider an additive fault f coupled to the input. Therefore, the model to be used is:

$$\dot{x}_1 = x_2 \tag{8.61}$$
$$\dot{x}_2 = \frac{1}{J}[c\phi x_3 - T_L]$$
$$\dot{x}_3 = \frac{1}{L_a}[-c\phi x_2 - R_a x_3 + u + f]$$
$$y = x_2.$$

8.4.3.1 Fault Diagnosis

From (8.61) the following polynomial is obtained:

$$f = c\phi y + R_a x_3 + L_a \dot{x}_3 - u \tag{8.62}$$

It can be seen that state x_3 must be also estimated somehow, thus it must also be algebraically observable. From (8.61):

$$x_3 = \frac{1}{c\phi}[J\dot{y} + T_L] \tag{8.63}$$

hence, state x_3 is algebraically observable and can also be reconstructed. Then, the fault is diagnosable.

In order to perform a good comparison with the fractional-order case, in this section the following integer-order FIROO will be used for FD:

$$\dot{\hat{f}}_i = k_{i0}(f_i - \hat{f}_i) + k_{i1}I(f_i - \hat{f}_i) \tag{8.64}$$

where I denotes an integer-order integration.

Therefore, we proceed to design the PIROO to estimate the fault:

$$\dot{\hat{f}} = k_{10}(f - \hat{f}) + k_{11}I(f - \hat{f}) \tag{8.65}$$
$$= k_{10}(c\phi y + R_a \hat{x}_3 + L_a \dot{\hat{x}}_3 - u - \hat{f}) + k_{11}I(c\phi y + R_a \hat{x}_3 + L_a \dot{\hat{x}}_3 - u - \hat{f})$$

In order to eliminate the derivative of $\hat{x}_3$ in the proporcional term, the auxiliary variable γ_1 is proposed and the estimation of the fault is defined as:

$$\hat{f} = \gamma_1 + k_{10}L_a \hat{x}_3 \tag{8.66}$$

and:

$$\dot{\gamma}_1 = \dot{\hat{f}} - k_{10}L_a \dot{\hat{x}}_3 \tag{8.67}$$
$$= k_{10}(c\phi y + R_a \hat{x}_3 + L_a \dot{\hat{x}}_3 - u - \hat{f} + k_{11}I(c\phi y + R_a \hat{x}_3 + L_a \dot{\hat{x}}_3 - u - \hat{f}) - k_{10}L_a \dot{\hat{x}}_3.$$

Therefore, the PIROO that will be used to estimate the fault is:

$$\dot{\gamma}_1 = k_{10}(c\phi y + R_a \hat{x}_3 - u - \gamma_1 - k_{10}L_a \hat{x}_3) + k_{11}L_a \hat{x}_3 \tag{8.68}$$
$$+ k_{11}I(c\phi y + R_a \hat{x}_3 - u - \gamma_1 - k_{10}L_a \hat{x}_3)$$
$$\hat{f} = \gamma_1 + k_{10}L_a \hat{x}_3$$

where $\hat{x}_3$ is obtained with:

$$\hat{x}_3 = \frac{1}{c\phi}[J\dot{y} + T_L]. \tag{8.69}$$

In order to estimate the derivative of y in (8.50), define $\xi = \dot{y}$, and the following PIROO is proposed:

$$\dot{\hat{\xi}} = k_{\xi 0}(\xi - \hat{\xi}) + k_{\xi 1}I(\xi - \hat{\xi}) \tag{8.70}$$
$$= k_{\xi 0}(\dot{y} - \hat{\xi}) + k_{\xi 1}I(\dot{y} - \hat{\xi})$$

Now, the auxiliary variable is introduced γ_ξ and the estimated variable $\hat{\xi}$ is defined as:

$$\hat{\xi} = \gamma_\xi + k_{\xi 0}y$$

Then:

$$\dot{\gamma}_\xi = \dot{\hat{\xi}} - k_{\xi 0}\dot{y}$$
$$= -k_{\xi 0}\hat{\xi} + k_{\xi 1}I(\dot{y} - \hat{\xi})$$
$$= k_{\xi 0}\left(-\gamma_\xi - k_{\xi 0}y\right) + k_{\xi 1}I(\dot{y} - \gamma_\xi - k_{\xi 0}y).$$

Therefore, the estimation of $\hat{x}_3$ is obtained with:

$$\hat{x}_3 = \frac{1}{c\phi}\left[J\hat{\xi} + T_L\right]$$ (8.71)

where $\hat{\xi}$ is obtained with the following PIROO:

$$\dot{\gamma}_\xi = -k_{\xi 0}\left(\gamma_\xi + k_{\xi 0}y\right) + k_{\xi 1}y - k_{\xi 1}I\left(\gamma_\xi + k_{\xi 0}y\right)$$ (8.72)
$$\hat{\xi} = \gamma_\xi + k_{\xi 0}y.$$

8.4.3.2 Fault-Tolerant Control

In order to apply the FTC, we must find the $\mathcal{FCOGM}$, for which we must first calculate the n derivatives of the output:

$$\dot{y} = \frac{1}{J}\left[c\phi x_3 - T_L\right]$$ (8.73)

$$\ddot{y} = \frac{c\phi}{J}\dot{x}_3 = \frac{c\phi}{L_a J}\left[-c\phi x_2 - R_a x_3 + u + f\right]$$

$$\dddot{y} = \frac{c\phi}{L_a J}\left[-c\phi \dot{x}_2 - R_a \dot{x}_3 + \dot{u} + \dot{f}\right]$$

$$= \frac{c\phi}{L_a J}\left[R_a c\phi y - c\phi \dot{y} + R_a^2 x_3 - R_a u - R_a f + \dot{u} + \dot{f}\right].$$

Now, a canonical form of the tracking error is created:

$$e_1 = y - y_R$$ (8.74)
$$\dot{e}_1 = \dot{y} - \dot{y}_R = e_2$$
$$\dot{e}_2 = \ddot{e}_1 = \ddot{y} - \ddot{y}_R = e_3$$
$$\dot{e}_3 = \dddot{e}_1 = \dddot{y} - \dddot{y}_R$$

and thus a HGO can be built for it:

$$\dot{\hat{e}}_1 = \hat{e}_2 - 3\theta\left(\hat{e}_1 - e_1\right)$$ (8.75)
$$\dot{\hat{e}}_2 = \hat{e}_3 - 3\theta^2\left(\hat{e}_1 - e_1\right)$$
$$\dot{\hat{e}}_3 = \frac{c\phi}{L_a J}\left[R_a c\phi y - c\phi \dot{y} + R_a^2 \hat{x}_3 - R_a \hat{u} - R_a \hat{f} + \dot{\hat{u}} + \dot{\hat{f}}\right] - \dddot{y}_R - \theta^3\left(\hat{e}_1 - e_1\right)$$

$$= -\sum_{i=1}^{3} a_i \hat{e}_i.$$

From (8.75), the following dynamics of the fault-tolerant controller is obtained:

$$\dot{\hat{u}} = \frac{L_a J}{c\phi} \left(-a_1\hat{e}_1 - a_2\hat{e}_2 - a_3\hat{e}_3 + \dddot{y}_R \right) - R_a c\phi y + c\phi\dot{y} \qquad (8.76)$$

$$- R_a^2 \hat{x}_3 + R_a\hat{u} + R_a\hat{f} - \dot{\hat{f}}.$$

This is the controller that will eliminate the effects of the fault in the system; however, it can be seen that an estimation of the derivative of y is also needed, hence the one obtained with (8.72) is used. Therefore, the dynamics of the fault-tolerant controller to be used is:

$$\dot{\hat{u}} = \frac{L_a J}{c\phi}(-a_1\hat{e}_1 - a_2\hat{e}_2 - a_3\hat{e}_3 + \dddot{y}_R) - R_a c\phi y + c\phi\hat{\xi} \qquad (8.77)$$

$$- R_a^2 \hat{x}_3 + R_a\hat{u} + R_a\hat{f} - \dot{\hat{f}}.$$

Thus, defining $\hat{u} = \hat{u}_1$ and $\hat{f} = \hat{f}_1$, the chain of integrators of the closed-loop system is:

$$\dot{\hat{e}}_1 = \hat{e}_2 - 3\theta \left(\hat{e}_1 - e_1 \right)$$
$$\dot{\hat{e}}_2 = \hat{e}_3 - 3\theta^2 \left(\hat{e}_1 - e_1 \right)$$
$$\dot{\hat{e}}_3 = \frac{c\phi}{L_a J}[R_a c\phi y - c\phi\hat{\xi} + R_a^2\hat{x}_3 - R_a\hat{u} - R_a\hat{f} + \dot{\hat{u}} + \dot{\hat{f}}] - \dddot{y}_R - \theta^3 \left(\hat{e}_1 - e_1 \right)$$
$$\dot{\hat{u}}_1 = \frac{L_a J}{c\phi}(-a_1\hat{e}_1 - a_2\hat{e}_2 - a_3\hat{e}_3 + \dddot{y}_R) - R_a c\phi y + c\phi\hat{\xi} - R_a^2\hat{x}_3 + R_a\hat{u}_1 + R_a\hat{f}_1 - \dot{\hat{f}}_1$$
$$\dot{\hat{f}}_1 = k_{10}(c\phi y + R_a\hat{x}_3 - u - \gamma_1 - k_{10}L_a\hat{x}_3) + k_{11}L_a\hat{x}_3$$
$$+ k_{11}I(c\phi y + R_a\hat{x}_3 - u - \gamma_1 - k_{10}L_a\hat{x}_3)$$
$$+ \frac{k_{10}L_a}{c\phi}\left[J(-k_{\xi 0}\left(\gamma_\xi + k_{\xi 0}y \right) + k_{\xi 1}y - k_{\xi 1}I(\gamma_\xi + k_{\xi 0}y) + k_{\xi 0}\hat{\xi}) + T_L \right].$$

8.4.3.3 Simulation Results

Simulations for this system were made in MATLAB Simulink for 20 s. The following values for the parameters were used:

R_a	$2.13 \, \Omega$
L_a	$0.00484 \, \text{H}$
$c\phi$	$0.0683 \, \text{Vs}$
J	$0.0001148 \, \text{kgm}^2$
T_L	$0.0608 \, \text{Nm}$
V_a	$12 \, \text{V}$

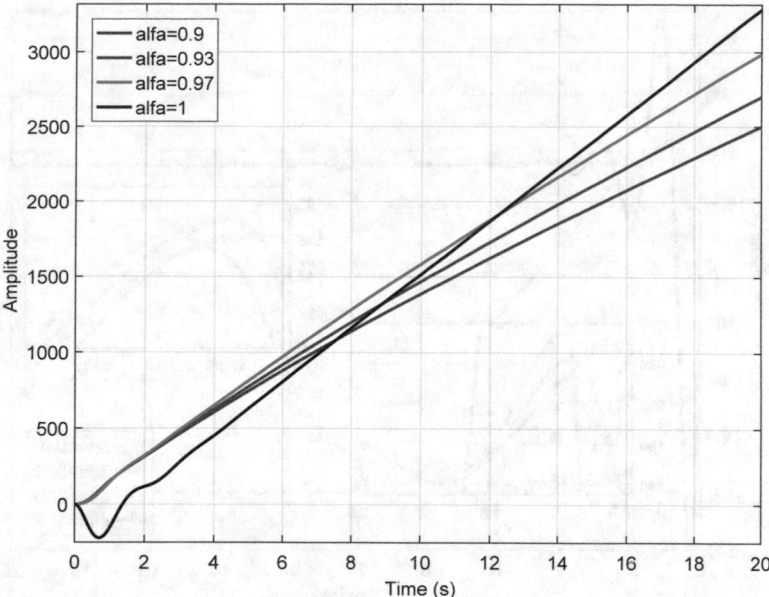

Fig. 8.11 Comparison of state x_1 of the DC motor between the integer and fractional-order cases

The reference was adjusted in $y_R = 177$. The fault was chosen as $f = 0.1V_a$, appearing at 10 s. The dimensionless design parameters (gains) were chosen as $\theta = 2000$, $a_1 = 8000$, $a_2 = 1200$, $a_3 = 60$, $k_{10} = k_{\xi 0} = 20$ and $k_{11} = k_{\xi 1} = 100$.

Same as in the fractional-order case, FD as well as OT were successful, thus the comparison of the results obtained will be done. For comparison purposes, simulations of the fractional-order model were performed with the values of α of 0.9, 0.93 and 0.97, being the integer-order case equivalent to the value of $\alpha = 1$.

Figure 8.11 compares state x_1 in the four cases, which corresponds to the angular position in the integer-order case and the α-order fractional integral of the angular velocity in the fractional-order case. It can be seen that in the integer-order case a bigger oscillation appears at the beginning, which leads to a straight line with a bigger slope; meanwhile, in the fractional-order case a more uniform curve with a smaller slope appears, which gets bigger and closer to the integer-order curve when the value of α increases.

Figure 8.12 compares output, i.e. state x_2 in the four cases, which corresponds to the angular velocity in all of them. It can be seen that, although all signals stabilize in a nearby value even in the occurrence of the fault at 10 s, the integer-order case has in the transient part an overshoot with a very big amplitude, which stabilizes in a bigger time. Similarly, it can be seen that in the fractional-order system the overshoot increases when the value of α gets closer to the unit.

Figure 8.13 compares state x_3 in the four cases, which corresponds to the armature current in all of them. Similarly to what happens with the output, the integer-order

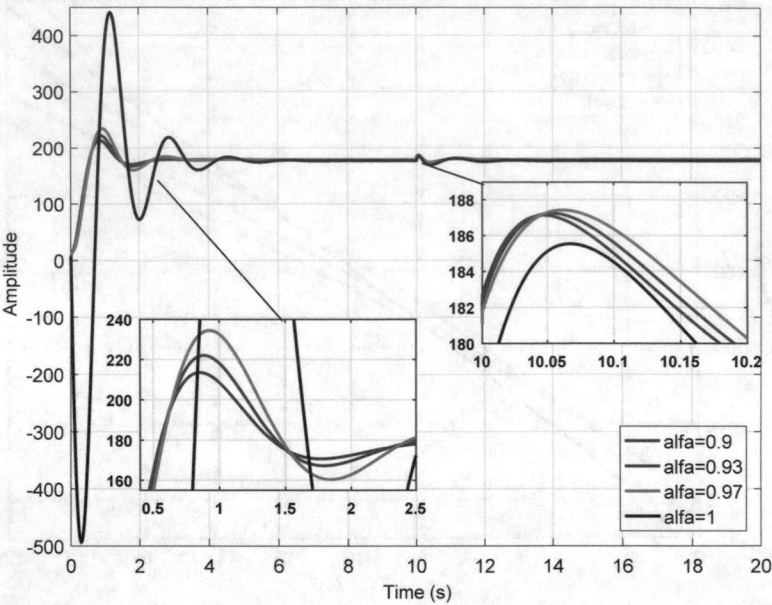

Fig. 8.12 Comparison of the output (state x_2) of the DC motor between the integer and fractional-order cases

case presents a transient part with bigger overshoot and settling time, coinciding the signals in both cases in the steady part; also, in the fractional-order case the overshoot increases when the value of α increases.

Figure 8.14 shows the comparison of the fault estimation $\hat{f}$ in the four cases. In a similar way to what has been observed, in the integer-order case a bigger transient overshoot appears at the beginning (without fault); when the fault appears, in all the cases an overshoot appears which stabilizes in a short period of time, but that is bigger when the value of α is bigger.

Figure 8.15 compares the FD performance index in the four cases. The index was calculated with the following cost function

$$J_t = \frac{1}{t + \varepsilon} \int_0^t \left\| \tilde{f} \right\|^2 dt$$

where $\varepsilon = 0.0001$. Due to the oscillating behaviour in the integer-order case it is appreciated that, although in all cases the index tends to zero, it is bigger in the integer-order case, and its amplitude diminishes while the value of α decreases.

Finally, Fig. 8.16 shows the comparison of the dynamical FTC signal $\hat{u}$ in the four cases. As it can be seen, the behaviour in the integer-order case shows bigger oscillations and settling time than in the fractional-order cases, on which a bigger overshoot is seen when the value of α is bigger.

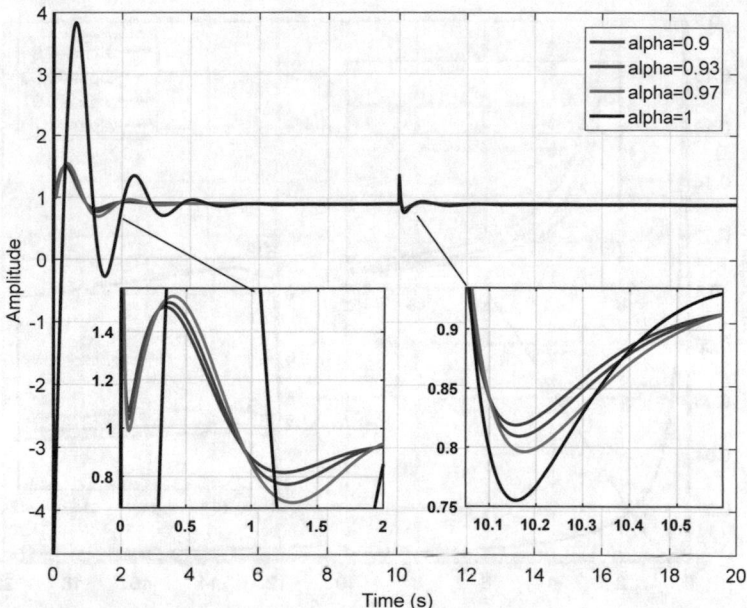

Fig. 8.13 Comparison of state x_3 of the DC motor between the integer and fractional-order cases

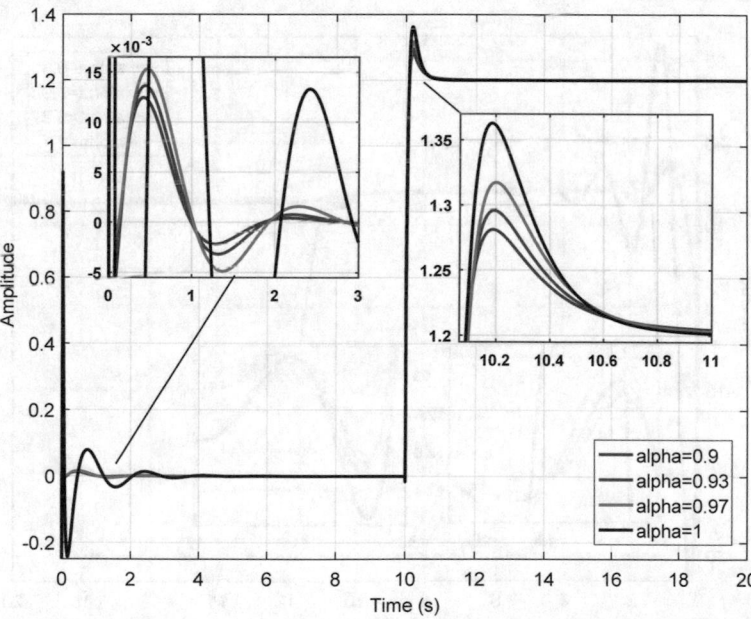

Fig. 8.14 Comparison of the fault estimation $\hat{f}$ of the DC motor between the integer and fractional-order cases

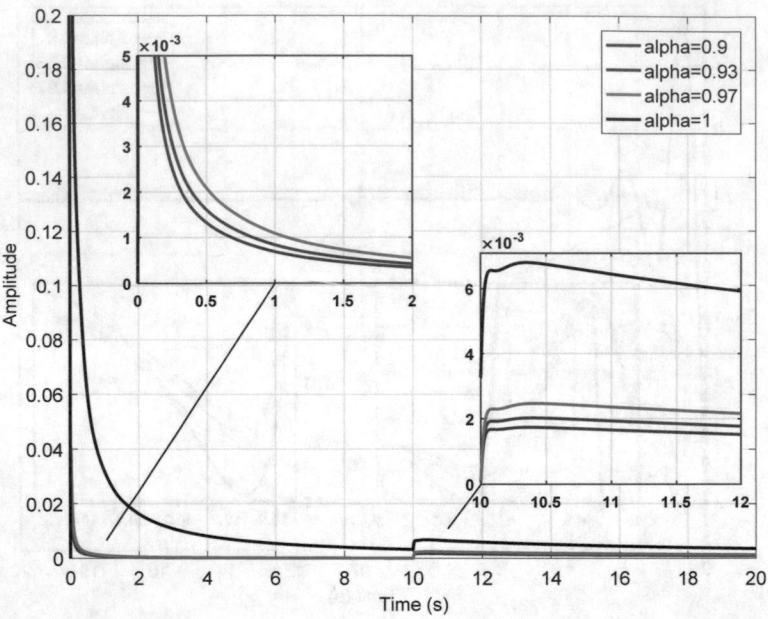

Fig. 8.15 Comparison of the FD performance index $\hat{f}$ of the DC motor between the integer and fractional-order cases

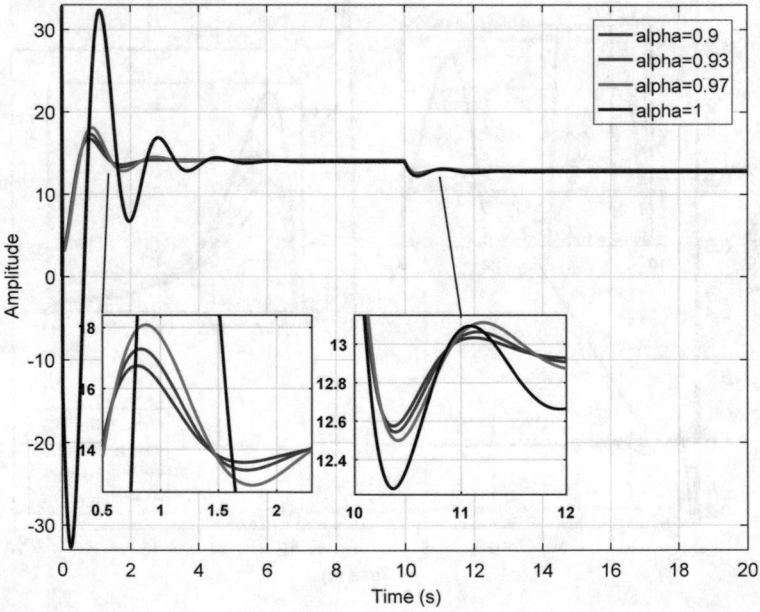

Fig. 8.16 Comparison of the dynamical FTC $\hat{u}$ of the DC motor between the integer and fractional-order cases

From these results, it can be concluded that with the use of fractional-order dynamics the model of the DC motor simulates to have certain kind of damping, which decreases when the value of α increases. This is why in these cases the signals, the angular velocity as well as the armature current, present small oscillations that stabilize in a short amount of time, as opposed to the results obtained with the integer-order dynamics. This could seem an advantage, because it can be concluded that the system behaves better with the fractional-order dynamics; however, it is important to consider how to perform the implementation of a model of this kind, besides of the physical meaning of the variables and dynamics involved. Instead of considering it as a better option, the fractional-order model can be taken together with a control system as an alternative to the classic controllers, in order to get different responses with a desired behaviour, different to the existing ones.

Furthermore, it can be appreciated the effect of the integral term of the PIROO in the model with integer-order dynamics, which causes overshoots of large magnitude to appear. It is worth to note how in the fractional-order cases the presence of an integral term does not increase in a big proportion the overshoot (see Figs. 8.1 and 8.2) and, although it may represent a better option for some systems, it must be taken into account also the numerical method used to implement the fractional operators, because it may lead to different results.

8.5 Concluding Remarks

In this chapter, a FTC scheme was proposed for a class of commensurate-order fractional nonlinear systems. This FTC scheme is hybrid, in the sense that it consists of a fractional proportional integral reduced-order observer and a fractional high-gain observer. A fractional fault-tolerant dynamical controller is designed in order to achieve output tracking in the presence of faults; the controller is obtained in a natural way from a Fractional Generalized Observability Canonical Form of the output tracking error. It is worth to mention that this controller uses estimations of the faults to eliminate simultaneously their effects. It is verified that the origin of the system in closed-loop with the fractional dynamical controller is Mittag-Leffler stable. The proposed methodology is assessed in the commensurate fractional models of the Van der Pol oscillator and a DC motor, where it can be seen by means of the performance indices that the faults are estimated successfully, and thus this leads to output tracking in each system.

References

1. Alcorta-García, E., Frank, P.: Deterministic nonlinear observer-based approaches to fault diagnosis: a survey. Control Eng. Pract. **5**(5), 663–670 (1997)
2. Aribi, A., Aoun, M., Farges, C., Najar, S., Melchior, P., Abdelkrim, M.N.: Generalized fractional observers scheme to fault detection and isolation. In: 10th International Multi-Conference on Systems, Signals & Devices (SSD), Hammamet, Tunisia, 18–21 March, pp. 1–7 (2013)
3. Aribi, A., Farges, C., Aoun, M., Melchior, P., Najar, S., Abdelkrim, M.N.: Fault detection based on fractional order models: application to diagnosis of thermal systems. Commun. Nonlinear Sci. **19**(10), 3679–3693 (2014)
4. Bao, H., Cao, J.: Existence of solutions for fractional stochastic impulsive neutral functional differential equations with infinite delay. Adv. Differ. Equ.-NY **66**(2017), 1–14 (2017)
5. Benchohra, M., Souid, M.S.: L1-solutions of boundary value problems for implicit fractional order differential equations. Surv. Math. Appl. **10**(2015), 49–59 (2015)
6. Blanke, M., Kinnaert, M., Lunze, J., Staroswiecki, M.: Diagnosis and Fault-Tolerant Control. Springer, Berlin (2003)
7. Chouki, R., Aribi, A., Aoun, M., Abdelkrim, M.N.: Additive fault tolerant control for fractional order model systems. In: Proceedings of the 16th International Conference on Sciences and Techniques of Automatic Control & Computer Engineering - STA'2015, Monastir, Tunisia, 21–23 December, pp. 340–345 (2015)
8. Cipin, R., Ondrusek, C., Huzlík, R.: Fractional-Order Model of DC Motor. In: Mechatronics 2013: Recent Technological and Scientific Advances. Springer, London (2013)
9. Cruz-Victoria, J.C., Martínez-Guerra, R., Pérez-Pinacho, C.A., Gómez-Cortés, G.C.: Synchronization of nonlinear fractional order systems by means of $PI^{r^{\alpha}}$ reduced order observer. Appl. Math. Comput. **262**(C), 224–231 (2015)
10. De Espíndola, J.J., da Silva Neto, J.M., Lopes, E.M.O.: A generalized fractional derivative approach to viscoelastic material properties measurement. Appl. Math. Comput. **164**(2), 493–506 (2005)
11. Fekih, A.: Fault-tolerant flight control design for effective and reliable aircraft systems. J. Control Decis. **1**(4), 299–316 (2014)
12. Gabano, J.D., Poinot, T.: Fractional modelling and identification of thermal systems. Signal Process **91**(3), 531–541 (2011)
13. Gao, Z.: Modulating function-based system identification for a fractional-order system with a time delay involving measurement noise using least-squares method. Int. J. Syst. Sci. **48**(7), 1460–1471 (2016)
14. Gaul, L., Klein, P., Kempfle, S.: Damping description involving fractional operators. Mech. Syst. Signal Pr. **5**(2), 81–88 (1991)
15. Gauthier, J.P., Hammouri, H., Othman, S.: A simple observer for nonlinear systems applications to bioreactors. IEEE Trans. Autom. Control **37**(6), 875–880 (1992)
16. Huang, C., Cao, J.: Active control strategy for synchronization and anti-synchronization of a fractional chaotic financial system. Phys. A **473**(2017), 262–275 (2017)
17. Jia, Q., Chen, W., Zhang, Y., Li, H.: Integrated design of fault reconstruction and fault-tolerant control against actuator faults using learning observers. Int. J. Syst. Sci. **47**(16), 3749–3761 (2016)
18. Kaczorek, T., Rogowski, K.: Fractional Linear Systems and Electrical Circuits. Springer, Switzerland (2015)
19. Liu, Y., Yang, G., Li, X.: Fault-tolerant control for uncertain linear systems via adaptive and LMI approaches. Int. J. Syst. Sci. **48**(2), 347–356 (2017)
20. Magin, R.: Fractional Calculus in Bioengineering. Begell House, Redding (2006)
21. Martínez-Guerra, R., Mata-Machuca, J.L.: Fault Detection and Diagnosis in Nonlinear Systems: A Differential and Algebraic Viewpoint. Springer, Cham (2014)
22. Martínez-Guerra, R., Mata-Machuca, J.L.: Fractional generalized synchronization in a class of nonlinear fractional order systems. Nonlinear Dynam. **77**(4), 1237–1244 (2014)

23. Metzler, R., Schick, W., Kilian, H.G., Nonnenmacher, T.F.: Relaxation in filled polymers: a fractional calculus approach. J. Chem. Phys. **103**(16), 7180–7186 (1995)

24. Mohajerpoor, R., Abdi, H., Nahavandi, S.: Reduced-order functional observers with application to partial state estimation of linear systems with input-delays. J. Control Decis. **2**(4), 233–256 (2015)

25. Monje, C.A., Chen, Y., Vinagre, B.M., Xue, D., Feliu, V.: Fractional-Order Systems and Controls: Fundamentals and Applications. Springer, London (2010)

26. Nieto, J.J., Ouahab, A., Venktesh, V.: Implicit fractional differential equations via the Liouville-Caputo derivative. Mathematics **3**(2015), 398–411 (2015)

27. Oldham, K.B., Spanier, J.: The Fractional Calculus: Theory and Applications of Differentiation and Integration to Arbitrary Order. Academic, New York (1974)

28. Patton, R.J.: Fault-tolerant control systems: the 1997 situation. In: IFAC Symposium on Fault Detection Supervision and Safety for Technical Processes, vol. 3, pp. 1033–1054 (1997)

29. Petráš, I.: Fractional-Order Nonlinear Systems: Modeling. Analysis and Simulation, Springer, Beijing (2011)

30. Pisano, A., Usai, E.: Second-order sliding mode approaches to disturbance estimation and fault detection in fractional-order systems. In: Proceedings of the 18th IFAC World Congress, Milano, Italy, 28 August–2 September, pp. 1033–1054 (2011)

31. Podlubny, I.: Fractional Differential Equations: An Introduction to Fractional Derivatives, Fractional Differential Equations, to Methods of their Solution and Some of their Applications. Academic, San Diego (1999)

32. Rosales, J.J., Gómez, J.F., Guía, M., Tkach, V.I.: Fractional electromagnetic waves. In: Proceedings of the LFNM*2011 International Conference on Laser & Fiber-Optical Networks Modeling, Kharkov, Ukraine, 4–8 September, pp. 1–3 (2011)

33. Sabatier, J., Oustaloup, A., García-Iturricha, A., Lanusse, P.: CRONE control: principles and extension to time-variant plants with asymptotically constant coefficients. Nonlinear Dynam. **29**(1), 363–385 (2002)

34. Scalas, E., Gorenflo, R., Mainardi, F.: Fractional calculus and continuous-time finance. Phys. A 284-384 (2000)

35. Shaw, S., Warby, M.K., Whiteman, J.R.: A comparison of hereditary integral and internal variable approaches to numerical linear solid elasticity. In: Proceedings of the XIII Polish Conference on Computer Methods in Mechanics, Pozna, Poland, 5–8 May (1997)

36. Shen, H., Song, X., Wang, Z.: Robust fault-tolerant control of uncertain fractional-order systems against actuator faults. IET Control Theory A **7**(9), 1233–1241 (2013)

37. Sun, Y., Wu, X., Cao, J., Wei, Z., Sun, G.: Fractional extended Kalman filtering for nonlinear fractional system with Lvy noises. IET Control Theory A **11**(3), 349–358 (2017)

38. Talange, D., Joshi, S.: Fractional order fault tolerant controller for AUV. In: Proceedings of the 18th International Conference on Automatic Control, Modelling & Simulation (ACMOS '16), Venice, Italy, 29–31 January, pp. 287–292 (2016)

39. Tavakoli-Kakhki, M.: Implementation of fractional-order transfer functions in the viewpoint of the required fractional-order capacitors. Int. J. Syst. Sci. **48**(1), 63–73 (2017)

40. Tidke, H.L., Mahajan, R.P.: Existence and uniqueness of nonlinear implicit fractional differential equation with Riemann-Liouville derivative. Amer. J. Comput Appl. Math. **7**(2), 46–50 (2017)

41. Wei, Y., Sun, Z., Hu, Y., Wang, Y.: On line parameter estimation based on gradient algorithm for fractional order systems. J Control Decis. **2**(4), 219–232 (2015)

42. Willsky, A.: A survey of design methods in observer-based fault detection systems. Automatica **1**(2), 601–611 (1976)

43. Xiao, M., Zheng, W.X., Cao, J.: Approximate expressions of a fractional order Van der Pol oscillator by the residue harmonic balance method. Math. Comput. Simul. **89**(2013), 1–12 (2013)

44. Yu, W., Luo, Y., Pi, Y.: Fractional order modeling and control for permanent magnet synchronous motor velocity servo system. Mechatronics **23**(7), 813–820 (2013)

Chapter 9
Fractional-Order Controller Based on a Robust PI$^\alpha$ Observer for Uncertain Fractional-Order Systems

Fractional calculus (FC) is an important branch of applied mathematics [15, 31, 33]; no matter how abstract the fractional operators look like, these can be used to describe elegantly physical phenomena. The fractional operators have exhibited the capability to represent accurately processes and they can give place to characterize dynamic systems providing a different perspective to understand the real world [10, 15, 22, 29, 32, 46].

The research on FC together with control theory have been gained attention in the community of automatic control; both topics present interesting tools to analyze fractional-order systems (FOS). In 1991, Oustaloup motivated the application of the FC in control theory by introducing a non-integer robust control scheme called CRONE [27, 28]. Since then, the control theory based on FC has been extensively studied. The classical PID controller based on fractional operators can be studied in various summaries (see [34, 43] and references therein). Furthermore, some practical applications like the control of congestion in TCP networks using fractional PID controllers can be found in literature [9]. Recently, a new method of fractional distributed dynamic matrix controller for fractional multivariable complex processes is presented in [40]. Meanwhile, a new methodology has been shown to obtain the robust stability region of fractional PI controllers for a fractional linear time-invariant plant [20]. The differential-algebraic technique is adopted in [24] to face the issue of synchronization as an observation problem. Another interesting approach for linear control applied to robust path tracking is the fractional flatness principle [38]. Newly in [26], an observer-based fault-tolerant control scheme is proposed for a class of commensurate fractional nonlinear systems. They presented a novel closed-loop analysis to demonstrate Mittag-Leffler stability.

Furthermore, the research on sliding mode control (SMC) lays down that it is a potent technique to control uncertain dynamical systems, with the cost of harmful chattering that risks real applications [36]. Moreover, the FC has been successfully

© The Author(s), under exclusive license to Springer Nature Switzerland AG 2021 165
R. Martínez-Guerra et al., *Fault-tolerant Control and Diagnosis for Integer and Fractional-order Systems*, Studies in Systems, Decision and Control 328,
https://doi.org/10.1007/978-3-030-62094-3_9

merged with sliding mode techniques [1, 2, 14, 21, 30, 39]. Previous studies on the sliding mode observer (SMO) have illustrated that the state estimation is useful to confront problems like synchronization and state feedback control, see [8, 24, 44, 45]. Several fractional observer-based control approaches for FOS have been presented in the literature in recent years. Observer-based robust control for fractional complex dynamic networks is introduced in [16]. The authors [11] consider an LMI approach to ensure the asymptotic stability of the fractional linear systems with parametric uncertainty. A study of the global asymptotic stability and observer-based stabilization problems of a class of fractional system is carried out in [17, 18]. The authors of [42] face the problem of general output feedback stabilization for fractional order linear time-invariant for $0 < \alpha < 2$, the objective is to design suitable output feedback controllers that guarantee the stability of the resulting closed-loop systems. The contribution provided by the authors [13] focuses on the state and static output feedback stabilization for fractional-order singular uncertain linear systems, the goal is to design suitable feedback controllers that guarantee the stability of resulting closed-loop control systems in terms of LMIs. In [6] is investigated the design of fractional surface for sliding mode control of a class of uncertain nonlinear system by LMI technique. The switching law is achieved guaranteeing the reachability condition, and the control law is established to obtain a sliding mode controller capable maintain the sliding motion. In [35] addresses the design of a robust fractional dynamic output-feedback sliding mode controller for a general class of uncertain systems subject to saturation element. The closed-loop system exhibits asymptotic stability, and the system states approach the sliding surface in a finite time.

In some control problems the fractional derivative of a signal is computed using numerical methods [12] and sometimes such method may not presents robustness against noise measurement; this fact attracted to the authors to study the issue of Robust Fractional Differentiator (RFD). Some applications of the RFD are image processing [4] and industrial fractional PID controllers [30, 41]. A key limitation of these research is that the algorithms for RFD are focused only on the estimation of the first α-derivative; moreover, if a fractional derivative is concatenated, then, the problem of the noise amplification arises.

The present chapter contributes to the theoretical study and application of fractional observers and fractional SMC for a class of FOS with bounded disturbance. The main contribution of this research is to propose a fractional sliding mode controller based on a PI^α observer to achieve closed-loop Mittag-Leffler stability. This methodology provides robustness against unknown input signals and chattering reduction. Because of the lack of measurable outputs in most practical problems, the proposed scheme is based on a fractional robust PI^α observer to provide the required information of state variables. This observer is designed such that a $\mathscr{L}_2$ stability is obtained via quadratic Lyapunov functions based on an LMI approach. Furthermore, with the purpose of avoiding concatenated fractional derivatives, a generalized $n\alpha$-differentiator which consists of a chain of integrators is introduced. A comparative analysis is performed to validate the methodology. The sliding mode controller is compared against the Ninteger toolbox [37] for Matlab® Simulink and the fractional super-twisting algorithm (FSTA) presented in [30]. The super-twisting algorithm

(STA) showed in [30] was designed to estimate only the α-derivative, due to this, the concatenation of the algorithm is needed to estimate high order derivatives (Fig. 9.2 presents a block diagram of the concatenation), a clear disadvantage compared with the authors proposal ($n\alpha$-differentiator).

9.1 Preliminaries

Some definitions and previous results are introduced to ease the understanding and support the proposed control scheme. The fractional operators used in the chapter and some useful theoretical results are presented.[1]

Definition 9.1 ([31]) The Euler gamma function $\Gamma(x)$ is defined by the integral

$$\Gamma(x) = \int_0^\infty t^{x-1} e^{-t} dt \tag{9.1}$$

which converges in the right half of the complex plane $Re(x) > 0$, with the property of $\Gamma(x+1) = x\Gamma(x)$, and for $n \in \mathbb{N}$, it yields $(n-1)! = \Gamma(n)$.

Definition 9.2 ([31]) A two-parameter function of the Mittag-Leffler type is defined by the series expansion

$$E_{\alpha,\beta}(z) = \sum_{k=0}^\infty \frac{z^k}{\Gamma(\alpha k + \beta)} \tag{9.2}$$

with $\alpha > 0$ and $\beta > 0$. The Mittag-Leffler function is a generalization of the exponential function; in particular, note that $E_\alpha(z) = E_{\alpha,1}(z)$, and $E_{1,1}(z) = e^z$.

Definition 9.3 ([15]) The Riemann–Liouville fractional integral of order α of a given signal $s(t)$ at instant time $t > 0$ is defined as

$$_0\mathscr{I}_t^\alpha s(t) = \frac{1}{\Gamma(\alpha)} \int_0^t \frac{s(t)}{(t-\tau)^{1-\alpha}} d\tau \tag{9.3}$$

where $Re(\alpha) > 0$, and $\Gamma(\cdot)$ denotes the Euler gamma function.

Definition 9.4 ([15, 31]) The Caputo fractional derivative of order $Re(\alpha) > 0$ of a given signal $s(t)$ at instant time $t \geq 0$ is defined as

---

[1]To present a less cumbersome notation, all fractional integral $\left(_0\mathscr{I}_t^\alpha\right)$ and derivatives $\left(_0^C\mathscr{D}_t^\alpha\right)$ are defined with lower limit equal to zero.

$$
{}_0^C \mathscr{D}_t^\alpha s(t) = {}_0\mathscr{I}_t^{n-\alpha} \left(\frac{d}{dt}\right)^n s(t) \tag{9.4}
$$

$$
= \frac{1}{\Gamma(\alpha - n)} \int_0^t \frac{s^{(n)}(\tau)\, d\tau}{(t-\tau)^{\alpha+1-n}}
$$

where $n - 1 < \alpha < n$, $n \in \mathbb{N}$, and $s^{(n)}(t)$ is the usual nth derivative of $s(t)$.

Lemma 9.1 ([15]) *Let $0 < \alpha < 1$ and $t \in [a, b]$. If $y(t) \in AC^n[a, b]$ or $y(x) \in C^n[a, b]$, then*

$$
\left({}_0\mathscr{I}_t^\alpha\, {}_0^C\mathscr{D}_t^\alpha y\right)(t) = y(t) - y(0). \tag{9.5}
$$

Corollary 9.1 ([7]) *For $a < b$, $\alpha > 0$ and $f \in C[a, b]$, there exists $\xi \in (a, b)$, then*

$$
{}_0\mathscr{I}_t^\alpha f(t) = \frac{1}{\Gamma(\alpha + 1)}(b - a)^\alpha f(\xi). \tag{9.6}
$$

Lemma 9.2 ([3]) *Let $x(t) \in \mathbb{R}^n$ be a continuous and derivable function. Then, for any time instant $t \geqslant t_0$:*

$$
\frac{1}{2}{}_{t_0}^C\mathscr{D}_t^\alpha x^2(t) \leqslant x(t)\, {}_{t_0}^C\mathscr{D}_t^\alpha x(t); \quad 0 < \alpha < 1.
$$

Consider the fractional-order system

$$
{}_{t_0}^C\mathscr{D}_t^\alpha x(t) = f(t, x) \tag{9.7}
$$

with initial condition $x(t_0)$, where $0 < \alpha < 1$, $f : [t_0, \infty] \times \Omega \to \mathbb{R}^n$ is piecewise continuous in t and locally Lipschitz in x on $[t_0, \infty] \times \Omega$, and $\Omega \in \mathbb{R}^n$ is a domain that contains the equilibrium point $x = 0$. The equilibrium point of (9.7) is defined as follows.

Definition 9.5 ([19]) The constant x_0 is an equilibrium point of the fractional order systems (9.7), if and only if ${}_{t_0}^C\mathscr{D}_t^\alpha x(t) = f(t, x_0)$.

Theorem 9.1 ([19]) *Let $x = 0$ be an equilibrium point for the system (9.7), and $\mathbb{D} \subset \mathbb{R}^n$ be a domain containing the origin. Let $V(t, x(t)) : [0, \infty) \times \mathbb{D} \to \mathbb{R}$ be a continuously differentiable function and locally Lipschitz with respect to x such that:*

$$
\alpha_1 \|x\|^a \leq V(t, x(t)) \leq \alpha_2 \|x\|^{ab}
$$
$$
{}_0^C\mathscr{D}_t^\beta V(t, x(t)) \leq -\alpha_3 \|x\|^{ab}
$$

where $t \geq 0$, $x \in \mathbb{D}$, $0 < \beta < 1$ and α_1, α_2, α_3, a and b are arbitrary positive constants. Then $x = 0$ is Mittag-Leffler stable. If the assumptions hold globally on $\mathbb{R}^n$, then $x = 0$ is globally Mittag-Leffler stable.

Remark 9.1 Mittag-Leffler stability implies asymptotic stability.

9.2 Robust PI$^\alpha$ Observer for Fractional Uncertain Systems

In this section, we address the problem of designing a robust PI$^\alpha$ observer for fractional systems with bounded disturbances. The observer has to ensure a robust state estimation independently of the uncertainty of the system which can be described by

$$^C\mathscr{D}^\alpha x(t) = Ax(t) + Bu(t) + H_\zeta \zeta(t), \qquad 0 < \alpha < 1 \tag{9.8}$$
$$y(t) = Cx(t)$$

where $x(t) \in \mathbb{R}^n$ is the state vector, $u(t) \in \mathbb{R}^l$ is the input vector, $y(t) \in \mathbb{R}^m$ is the output vector, $\zeta(t) \in \mathbb{R}^\mu$ is the disturbance vector, $A \in \mathbb{R}^{n \times n}$, $B \in \mathbb{R}^{n \times l}$, $C \in \mathbb{R}^{m \times n}$ and $H_\zeta \in \mathbb{R}^{n \times \mu}$ are known constant matrices.

The following assumption is needed through this chapter.

Assumption 9.1 The matrix pairs (A, B) and (A, C) are controllable and observable, respectively.

Let us consider the following robust PI$^\alpha$ observer for the class of fractional systems described by (9.8):

$$^C\mathscr{D}^\alpha \hat{x}(t) = A\hat{x}(t) + Bu(t) + K_p\left(y(t) - \hat{y}(t)\right) + K_i z(t)$$
$$^C\mathscr{D}^\alpha z(t) = A_z z(t) + y(t) - \hat{y}(t) \tag{9.9}$$
$$\hat{y}(t) = C\hat{x}(t)$$

where $K_p \in \mathbb{R}^{n \times m}$ and $K_i \in \mathbb{R}^{n \times m}$ are the proportional and integral gain respectively. The vector $z(t)$ represents the integral of the output estimation error whose time response can be modified using $A_z \in \mathbb{R}^{m \times m}$ which is an unknown diagonal exponentially stable matrix.

Remark 9.2 The gains K_p and K_i have to be tuned depending on the performance specification.

9.2.1 Stability Analysis

Let us begin defining the observation error $\tilde{x}(t) = x(t) - \hat{x}(t)$. Then, the observer error dynamic is defined by

$$^C\mathscr{D}^\alpha \tilde{x}(t) = A\tilde{x}(t) - K_p C\tilde{x}(t) - K_i z(t) + H_\zeta \zeta(t) \tag{9.10}$$
$$\tilde{y} = C\tilde{x}(t)$$

Now, the sufficient conditions for $\mathscr{L}_2$ stability via quadratic Lyapunov functions based on a linear matrix inequality approach are presented.

Theorem 9.2 *Consider the fractional system (9.8) and the PI^α observer (9.9) satisfying Assumption 9.1. If there exist some matrices $P = P^T > 0$, $Q = Q^T > 0$ and a constant $\gamma > 0$, such that*

$$
\begin{bmatrix}
A_x^T P + P A_x + C^T C & C^T Q - P K_i & P H_\zeta \\
Q C - K_i^T P & A_z^T Q + Q A_z & 0 \\
H_\zeta^T P & 0 & -\gamma^2 I
\end{bmatrix} < 0 \tag{9.11}
$$

Then, for given matrices A_z, K_p, K_i, and with $A_x = A - K_p C$ the observation error dynamic (9.10) is quadratically stable and have a $\mathscr{L}_2$ gain from ζ to $\tilde{y}$ which is smaller than γ.

Proof Let us consider the following quadratic Lyapunov function for the error dynamics described by (9.10)

$$
V(\tilde{x}, z) = \tilde{x}^T P \tilde{x} + z^T Q z \tag{9.12}
$$

where $P = P^T > 0$ and $Q = Q^T > 0$. The next step is to prove the stability of the observation error dynamic and to ensure that the $\mathscr{L}_2$ gain of the observer is no more than γ, which guarantees that all estimates are bounded. Then, the $\mathscr{L}_2$ is defined as:

$$
\|\tilde{y}\|_2 < \gamma \|\zeta\|_2 \tag{9.13}
$$

likewise, the inequality (9.13) can be rewritten as:

$$
\tilde{y}^T \tilde{y} - \gamma^2 \zeta^T \zeta < 0 \tag{9.14}
$$

Consequently, to check the quadratic stability condition of our robust PI^α observer and to bound the $\mathscr{L}_2$ gain of the error dynamics, we have to ensure that the inequality

$$
{}^C \mathscr{D}^\alpha V + \tilde{y}^T \tilde{y} - \gamma^2 \zeta^T \zeta < 0
$$

is satisfied.

Then, considering Lemma 9.2 and let us take the fractional derivative of (9.12) to obtain:

$$
{}^C \mathscr{D}^\alpha V \leq {}^C \mathscr{D}^\alpha V_{\tilde{x}} + {}^C \mathscr{D}^\alpha V_z \tag{9.15}
$$

where $V_{\tilde{x}} = \tilde{x}^T P \tilde{x}$, $V_z = z^T Q z$. The fractional derivative ${}^C \mathscr{D}^\alpha V_{\tilde{x}}$ is defined as

$$
{}^C \mathscr{D}^\alpha V_{\tilde{x}} \leq 2 \tilde{x}^T P \left[A \tilde{x} - K_p C \tilde{x} - K_i z + H_\zeta \zeta \right]
$$

$$
{}^C \mathscr{D}^\alpha V_{\tilde{x}} \leq \tilde{x}^T \left[(A - K_p C)^T P + P (A - K_p C) \right] \tilde{x} - 2 \tilde{x}^T P \left[K_i z - H_\zeta \zeta \right]
$$

and finally, $^C\mathscr{D}^\alpha V_z$ as:

$$^C\mathscr{D}^\alpha V_z \le 2z^T Q \left[A_Z z + C\tilde{x} \right]$$
$$^C\mathscr{D}^\alpha V_z \le \tilde{z}^T \left[A_z Q + Q A_z^T \right] \tilde{z} + 2z^T QC\tilde{x}$$

thus

$$^C\mathscr{D}^\alpha V \le \tilde{x}^T \left[\left(A - K_p C \right)^T P + P \left(A - K_p C \right) \right] \tilde{x} + \tilde{z}^T \left[A_z Q + Q A_z^T \right] \tilde{z}$$
$$+ \tilde{x}^T \left[(QC - K_i^T P) + (C^T Q - P K_i) \right] z + 2\tilde{x}^T P H_\xi \zeta$$

The fractional derivative of the Lyapunov function (9.12) is negative semi-definite if the following inequality holds.

$$\tilde{x}^T \left[\left(A - K_p C \right)^T P + P \left(A - K_p C \right) \right] \tilde{x} + \tilde{z}^T \left[A_z Q + Q A_z^T \right] \tilde{z}$$
$$+ \tilde{x}^T \left[(QC - K_i^T P) + (C^T Q - P K_i) \right] z + 2\tilde{x}^T P H_\xi \zeta$$
$$+ \tilde{y}^T \tilde{y} - \gamma^2 \zeta^T \zeta < 0 \qquad (9.16)$$

By introducing $A_x = A - K_p C$ and $\tilde{y}^T \tilde{y} = \tilde{x}^T C^T C \tilde{x}$ we can represent the above inequality as

$$\begin{bmatrix} x \\ z \\ \zeta \end{bmatrix}^T \begin{bmatrix} A_x^T P + P A_x + C^T C & C^T Q - P K_l & P H_\xi \\ QC - K_i^T P & A_z^T Q + Q A_z & 0 \\ H_\xi^T P & 0 & -\gamma^2 I \end{bmatrix} \begin{bmatrix} \tilde{x} \\ z \\ \zeta \end{bmatrix} < 0 \qquad (9.17)$$

and consequently, the inequality (9.17) can be represented as an equivalent LMI problem over P and Q as follows

$$\begin{bmatrix} A_x^T P + P A_x + C^T C & C^T Q - P K_i & P H_\xi \\ QC - K_i^T P & A_z^T Q + Q A_z & 0 \\ H_\xi^T P & 0 & -\gamma^2 I \end{bmatrix} < 0 \qquad (9.18)$$

then (9.12) is a Lyapunov function when (9.18) holds. Then, we can ensure the stability of our proposed observer by solving the LMI problem (9.18). $\qquad \square$

Due to our proposed methodology involves tuning several matrices, a method to compute the optimal parameters by reducing the effect of disturbances is given, that means minimizing the $\mathscr{L}_2$ gain in (9.11).

Theorem 9.3 *There exist optimal parameters K_p, K_i, and A_z that minimize the $\mathscr{L}_2$ gain if there exist the matrices $P = P^T > 0$, $Q = Q^T > 0$, R, S, $U = U^T < 0$ and the scalar $\gamma > 0$ such that*

$$\begin{bmatrix} \mathscr{A} - R^T C - C^T R & C^T Q - S & P H_\zeta \\ QC - S^T & 2U & 0 \\ H_\zeta^T P & 0 & -\gamma^2 I \end{bmatrix} < 0 \qquad (9.19)$$

with $\mathscr{A} = A^T P + P A + C^T C$.

Proof As a first step, let us define the matrices $R = K_p^T P$, $S = P K_i$ and $U = A_z Q$. Then we can rewrite the inequality (9.16) as

$$\tilde{x}^T \left[\mathscr{A} - R^T C - C^T R \right] \tilde{x} + \tilde{z}^T \left[U + U^T \right] \tilde{z}$$
$$+ \tilde{x}^T \left[(QC - S^T) + (C^T Q - S) \right] z + 2 \tilde{x}^T P H_\zeta \zeta$$
$$+ \tilde{y}^T \tilde{y} - \gamma^2 \zeta^T \zeta < 0 \qquad (9.20)$$

which can be straightforwardly rearranged as the minimization LMI problem

$$\begin{bmatrix} \mathscr{A} - R^T C - C^T R & C^T Q - S & P H_\zeta \\ QC - S^T & 2U & 0 \\ H_\zeta^T P & 0 & -\gamma^2 I \end{bmatrix} < 0 \qquad (9.21)$$

completing the proof. $\qquad\qquad\qquad\qquad\qquad\qquad\qquad\qquad\qquad\qquad\qquad\qquad\square$

Remark 9.3 Once the minimization problem is solved, the resulting optimal parameters can be computed as $K_p = P^{-1} R^T$, $K_i = P^{-1} S$ and $A_z = U Q^{-1}$.

9.3 Fractional-Order Sliding Mode Control

This section presents the closed-loop analysis of the fractional order linear system with matching uncertainty and a fractional order observer. It is proposed a dynamic control law based on the sliding mode technique to deal with uncertainty and to achieve the asymptotic convergence of the system state with the feedback of the information provided by the observer. Consider the fractional order linear system with matching uncertainty (9.8) and consider the fractional order observer (9.9).

Let be the following manifold:

$$\Sigma = \left\{ \hat{x} \in \mathbb{R}^n : \sigma \left(\hat{x}(t) \right) = 0 \right\} \qquad (9.22)$$

where σ is known as the sliding surface and it is defined as follows:

$$\sigma \left(\hat{x}(t) \right) = {}^C \mathscr{D}^\alpha w \left(\hat{x}(t) \right) + cw \left(\hat{x}(t) \right) \qquad (9.23)$$

with $w\left(\hat{x}\left(t\right)\right) = K\hat{x}$, $K \in \mathbb{R}^{l \times n}$ is a proportional gain, $c > 0$ is a positive constant, $0 < \alpha < 1$ is the fractional-order derivative.

Taking the fractional derivative of (9.23), it follows:

$$
\begin{aligned}
{}^{C}\mathscr{D}^{\alpha}\sigma &= {}^{C}\mathscr{D}^{2\alpha}w + c\,{}^{C}\mathscr{D}^{\alpha}w \\
&= K^{C}\mathscr{D}^{2\alpha}\hat{x} + cK^{C}\mathscr{D}^{\alpha}\hat{x} \\
&= KA\left[A\hat{x} + Bu + K_p\left(y - \hat{y}\right) + K_i z\right] + KB^{C}\mathscr{D}^{\alpha}u + KK_p{}^{C}\mathscr{D}^{\alpha}\left(y - \hat{y}\right) \\
&\quad + KK_i\left(A_z z + y - \hat{y}\right) + cK\left[A\hat{x} + Bu + K_p\left(y - \hat{y}\right) + K_i z\right] \\
&= K\left(A + cI\right){}^{C}\mathscr{D}^{\alpha}\hat{x} + KB^{C}\mathscr{D}^{\alpha}u + KK_p{}^{C}\mathscr{D}^{\alpha}\left(y - \hat{y}\right) + KK_i\left(A_z z + y - \hat{y}\right)
\end{aligned}
$$

Since the estimation error is defined as $\tilde{x}(t) = x(t) - \hat{x}(t)$, then $y - \hat{y} = Cx(t) - C\hat{x}(t) = C\tilde{x}(t)$, hence the fractional-order derivative of sliding surface remains as:

$$
{}^{C}\mathscr{D}^{\alpha}\sigma = K\left(A + cI\right){}^{C}\mathscr{D}^{\alpha}\hat{x} + KB^{C}\mathscr{D}^{\alpha}u + KK_p C^{C}\mathscr{D}^{\alpha}\tilde{x} + KK_i{}^{C}\mathscr{D}^{\alpha}z. \quad (9.24)
$$

9.3.1 Convergence Analysis

Lemma 9.3 *The fractional sliding surface* (9.23) *with a constant $c > 0$ is finite time reachable at $T > 0$ using the fractional dynamic control given by*

$$
{}^{C}\mathscr{D}^{\alpha}u = -\left(KB\right)^{-1}\left[K\left(A + cI\right){}^{C}\mathscr{D}^{\alpha}\hat{x} + KK_i{}^{C}\mathscr{D}^{\alpha}z + h\,sign\left(\sigma\right)\right]
$$

with $det\left(KB\right) \neq 0$ and $h = \gamma^{+} + h^{*}$ with $h^{*} > 0$.

Proof By considering the candidate Lyapunov function $V\left(t\right) = \frac{1}{2}\sigma^{T}\sigma$ and taking the fractional-order derivative, it yields:

$$
\begin{aligned}
{}^{C}\mathscr{D}^{\alpha}V &\leq \sigma^{T}\left({}^{C}\mathscr{D}^{\alpha}\sigma\right) \\
&\leq \sigma^{T}\left[K\left(A + cI\right){}^{C}\mathscr{D}^{\alpha}\hat{x} + KB^{C}\mathscr{D}^{\alpha}u + KK_p C^{C}\mathscr{D}^{\alpha}\tilde{x} + KK_i{}^{C}\mathscr{D}^{\alpha}z\right]
\end{aligned}
$$

with the fractional dynamic controller defined as:

$$
{}^{C}\mathscr{D}^{\alpha}u = -\left(KB\right)^{-1}\left[K\left(A + cI\right){}^{C}\mathscr{D}^{\alpha}\hat{x} + KK_i{}^{C}\mathscr{D}^{\alpha}z + h\,sign\left(\sigma\right)\right] \quad (9.25)
$$

The fractional derivative of $V(t)$ is reduced as follows:

$$
\begin{aligned}
{}^{C}\mathscr{D}^{\alpha}V &\leq \sigma^{T}\left[KK_p C^{C}\mathscr{D}^{\alpha}\tilde{x} - h\,sign\left(\sigma\right)\right] \\
&= \sigma^{T}KK_p C^{C}\mathscr{D}^{\alpha}\tilde{x} - \sigma^{T}h\,sign\left(\sigma\right) \\
&\leq \gamma^{+}\left|\sigma\right| - h\left|\sigma\right| \\
&\leq -\left(h - \gamma^{+}\right)\left|\sigma\right|
\end{aligned}
$$

with $h > \gamma^+$, the system state converges asymptotically to the origin, while $h > 0$ and $\gamma^+ > \left| K K_p C \tilde{x}^{(\alpha)} \right|$.

Finally, choosing $h = \gamma^+ + h^*$ with $h^* > 0$ it yields

$$
\begin{aligned}
{}^C\mathscr{D}^\alpha V(t) &\le -h^* |\sigma(t)| \\
&\le -\bar{h}^* \sqrt{V(t)} < 0
\end{aligned}
\tag{9.26}
$$

with $\bar{h}^* = h^* \sqrt{2}$.

Therefore, as the sufficient condition $\sigma(t)\mathscr{D}^\alpha \sigma(t) < 0$ is satisfied, the states variables of system (9.8) reach the sliding surface in a finite time defined as follows:

$$
\mathscr{I}^\alpha \left({}^C\mathscr{D}^\alpha V(t)\right) \le \mathscr{I}^\alpha \left(-\bar{h}^* \sqrt{V(t)}\right)
$$

Using the Lemma 9.1, we yield

$$
V(T) - V(0) \le -\frac{\bar{h}^*}{\Gamma(\alpha)} \int_0^T (t - \tau)^{\alpha-1} V^{1/2}(\tau)\, d\tau
$$

By applying the mean value theorem for the fractional integral (Corollary 9.1)

$$
V(T) - V(0) \le -\frac{\bar{h}^*}{\Gamma(\alpha + 1)} T^\alpha V^{1/2}(\xi)
$$

where $\xi \in [0, T]$. Then, in order to obtain the reaching time to the sliding surface, we substitute T by $t_r \in [0, T]$ as follows

$$
V(t_r) - V(0) \le -\frac{\bar{h}^*}{\Gamma(\alpha + 1)} t_r^\alpha V^{1/2}(\xi)
$$

Since $V(t_r) = 0$, we have the following

$$
t_r \le \sqrt[\alpha]{\frac{\alpha \Gamma(\alpha) V(0)}{\bar{h}^* V^{1/2}(\xi)}}
\tag{9.27}
$$

Since, the sliding variable $\sigma(t)$ reach the equilibrium point $\sigma(t) = 0$ at $t = t_r$, then for all $t \ge t_r$, the auxiliary variable $w(t)$ in (9.23) converges to the origin as follows:

$$
\begin{aligned}
{}^C\mathscr{D}^\alpha w(t) + cw(t) &= 0 \\
{}^C\mathscr{D}^\alpha w(t) &= -cw(t)
\end{aligned}
\tag{9.28}
$$

Solving the above Eq. (9.28), it results in

$$w(t) = E_\alpha(-ct^\alpha) w(0) \tag{9.29}$$

when $t \longrightarrow \infty$, then $E_{\alpha,1}(-ct^\alpha) \cdot w(0) \longrightarrow 0$, for all $c > 0$.

The inequality (9.26) confirms the existence of the sliding mode dynamics. Finally, based on Theorem 9.1; it can be concluded that the closed-loop system is Mittag-Leffler stable via the fractional sliding mode control, i.e, we have

$$^C\mathscr{D}^\alpha z(t) + \lambda z(t) = 0, \quad t > t_r > 0 \tag{9.30}$$

choosing $\lambda > 0$ the fractional dynamics (9.30) is asymptotically stable [15, 31], i.e,

$$\lim_{t\to\infty} |z(t)| = K \lim_{t\to\infty} |\hat{x}(t)| = 0.$$

$\square$

Remark 9.4 The sliding surface (9.23) restricted into the manifold (9.22) provides the following advantages; since the Lemma 9.3 proposes a dynamic control, the resulting signal $u(t)$ is continuous, hence the chattering due to the switching function is drastically reduced, accordingly, the input signal effort is reduced by the dynamic control and the matching perturbation is eliminated. Moreover, the sliding variable $\sigma(\hat{x}(t)) = 0$ is reached in finite-time and thereafter converge with Mittag-Leffler rate to the origin.

9.4 Case Study: Nα-Differentiator

The section introduces a generalized $n\alpha$-differentiator in order to validate the proposed methodology. Firstly, let us consider a signal $r(t) \in AC^n[a, b]$, where AC^n is the absolute continuous function space defined on the finite interval $[a, b]$ with $-\infty < a < b < \infty$, then, the function $r(t)$ is $n\alpha$th Caputo differentiable with $n \in \mathbb{N}$ and $n - 1 < \alpha \leq n$, see [15]. The $r(t)$ signal is supposed to fulfill the following smoothness restriction

Assumption 9.2 The $n\alpha$-derivative $^C\mathscr{D}^{n\alpha}r(t)$ is upper bounded, i.e,

$$|^C\mathscr{D}^{n\alpha}r(t)| \leq \gamma_r^+ < \infty. \tag{9.31}$$

9.4.1 Generalized Nα-Differentiator

Firstly, let us propose the structure of a $n\alpha$-order fractional differential equation of the measurable signal $r(t)$ as follows

$$\varepsilon(t) \triangleq r(t) - \hat{r}(t)$$
$$^C\mathcal{D}^{n\alpha}\hat{r}(t) = -u(t) \qquad\qquad 0 < \alpha < 1 \qquad\qquad (9.32)$$

where $\hat{r}(t)$ is the estimation signal, $\varepsilon(t)$ is called estimation error and $u(t)$ is an input signal or control signal.

With the purpose of building a generalized fractional canonical form (GFCF) for the system (9.32), the following set of state variables are defined as:

$$x_i(t) \triangleq {}^C\mathcal{D}^{(i-1)\alpha}\varepsilon(t) \qquad i = 1, 2, \ldots, n$$

Taking their αth time derivative, the following fractional dynamic system is obtained:

$$^C\mathcal{D}^\alpha x_i(t) = x_{(i+1)}(t), \qquad\quad i = 1, 2, \ldots, n-1$$
$$^C\mathcal{D}^\alpha x_n(t) = {}^C\mathcal{D}^{n\alpha}r(t) + u(t) \qquad\qquad (9.33)$$
$$y(t) = x_1(t)$$

The fractional system (9.33) can be rewritten in a matrix form described in (9.8) with the state vector $x(t) = [x_1(t), \ x_2(t), \ \ldots, \ x_n(t)]^T$, the control signal $u(t) \in \mathbb{R}$, the output of the system $y(t) \in \mathbb{R}$ and $\zeta(t) = {}^C\mathcal{D}^{n\alpha}r(t) \in \mathbb{R}$ as a matched perturbation which satisfies Eq. (9.31). The constant matrices $A \in \mathbb{R}^{n \times n}$, $B \in \mathbb{R}^{n \times 1}$ and $C \in \mathbb{R}^{1 \times n}$ are defined as:

$$A = \begin{bmatrix} 0 & 1 & 0 & \cdots & 0 \\ 0 & 0 & 1 & \cdots & 0 \\ 0 & 0 & 0 & \ddots & \vdots \\ \vdots & \vdots & \vdots & \ddots & 1 \\ 0 & 0 & 0 & \cdots & 0 \end{bmatrix}; \ B = \begin{bmatrix} 0 \\ 0 \\ \vdots \\ 1 \end{bmatrix}; \ C = \begin{bmatrix} 1 & 0 & \cdots & 0 \end{bmatrix} \qquad (9.34)$$

Figure 9.1 shows the closed-loop schematic for the generalized $n\alpha$-differentiator.

9.4.1.1 Numerical Results: 2α-Differentiator

Consider $n = 2$ and a measurable signal $r(t)$ that satisfies Assumption 9.2 and it is given by the following equation

$$r(t) = sin(t) + 0.2sin(2t)$$

Based on Eq. (9.9) and choosing $\alpha = 0.935$, the fractional observer for system (9.33) is described as

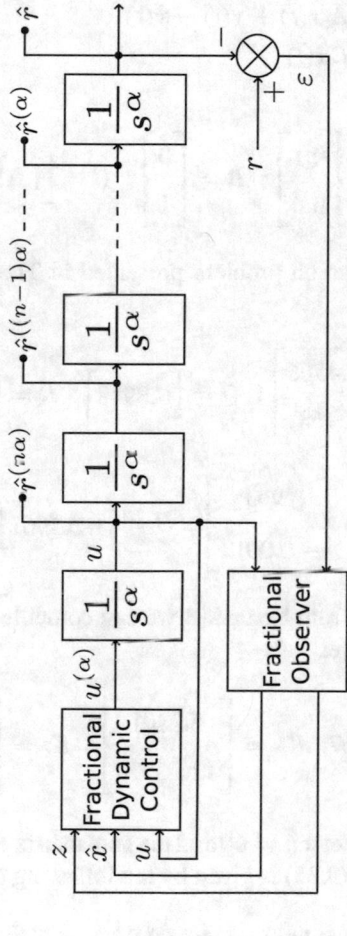

Fig. 9.1 Closed-loop schematic for the *nα*-differentiator

Fig. 9.2 Concatenated
fractional super-twisting
differentiator

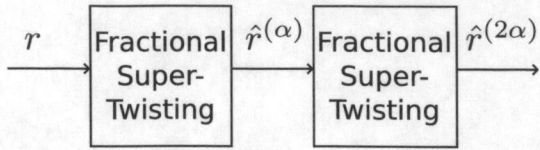

$$
{}^{C}\mathscr{D}^{0.935}\hat{x}(t) = A\hat{x}(t) + Bu(t) + K_p\left(y(t) - \hat{y}(t)\right) + K_i z(t)
$$
$$
{}^{C}\mathscr{D}^{0.935}z(t) = A_z z(t) + y(t) - \hat{y}(t) \tag{9.35}
$$
$$
\hat{y}(t) = C\hat{x}(t)
$$

with the matrices

$$
A = \begin{bmatrix} 0 & 1 \\ 0 & 0 \end{bmatrix}; \quad B = \begin{bmatrix} 0 \\ 1 \end{bmatrix}; \quad C = \begin{bmatrix} 1 & 0 \end{bmatrix}; \tag{9.36}
$$

By solving the minimization problem presented in Theorem 9.3 the following matrices are obtained

$$
P = \begin{bmatrix} 0.5349 & -0.4566 \\ -0.4566 & 0.5168 \end{bmatrix}; \quad Q = \begin{bmatrix} 0.8942 \end{bmatrix}; \quad R = \begin{bmatrix} 1.5323 & 0.3468 \end{bmatrix}
$$

$$
S = \begin{bmatrix} 0.8942 \\ -0.0012 \end{bmatrix}; \quad U = \begin{bmatrix} -0.5961 \end{bmatrix}
$$

and by using the relations from Remark 3 we can compute the optimal parameters for the proposed PI^{α} observer

$$
A_z = -0.6667; \quad K_p = \begin{bmatrix} 6.2801 \\ 126.3590 \end{bmatrix}; \quad K_i = \begin{bmatrix} 0.9329 \\ 3.9383 \end{bmatrix} \tag{9.37}
$$

Finally, taking the parameter $h = 60$ and the gain matrix $K = [10\ 1]^T$ the dynamical control for the system (9.35) is given by the following equation:

$$
{}^{C}\mathscr{D}^{0.935}u = -(KB)^{-1}\left[K(A + cI)\,{}^{C}\mathscr{D}^{0.935}\hat{x} + KK_i\,{}^{C}\mathscr{D}^{0.935}z + h\,sign\,(\sigma)\right] \tag{9.38}
$$

where $\sigma(t)$ is the fractional sliding surface and $c = 1$.

Figures 9.3, 9.4 and 9.5 illustrate the simulation results for the 2α-differentiator. The results are compared with the Matlab® Simulik Ninteger Toolbox and the FSTA [30]. Figure 9.2 shows the block diagram of the concatenated fractional super-twisting differentiator to estimate 2α-derivative of the signal $r(t)$.

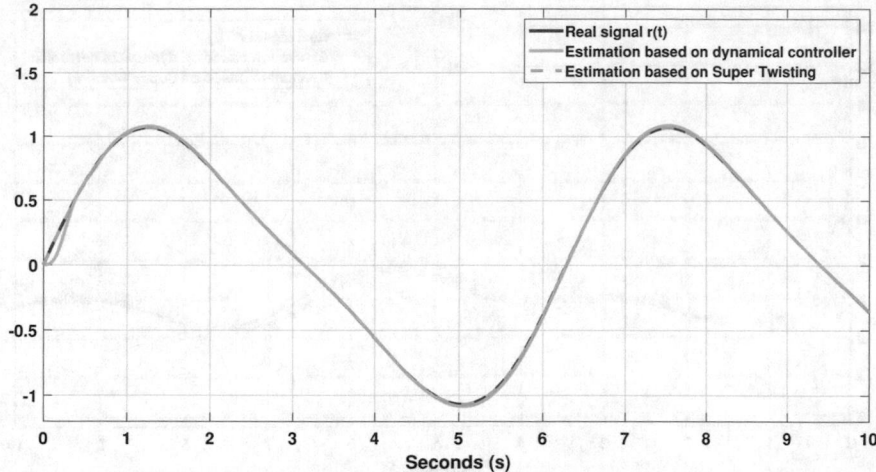

Fig. 9.3 Comparative analysis of the signal estimation $r(t)$ (noise free simulation)

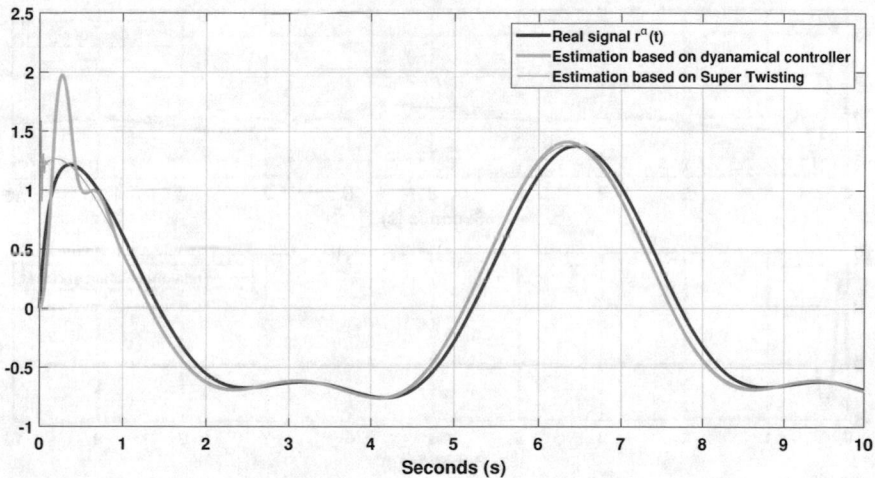

Fig. 9.4 Comparative analysis of the signal estimation $r^{(\alpha)}(t)$ (noise free simulation)

As we can see, our proposed controller has a filtering effect providing chattering reduction and tolerance for noisy measurement signals. Also, it is not necessary to concatenate our algorithm due to we present a generalization, another advantage over the STA. The fractional controller $u(t)$ and the fractional order sliding surface are displayed in Fig. 9.6.

With the purposes of evaluating and showing the robustness of the proposed methodology, the measurable signal $r(t)$ is contaminated using a Gaussian noise Δ_r, with zero mean, variance 0.001 and it is bounded within the interval $[-0.001\ 0.001]$.

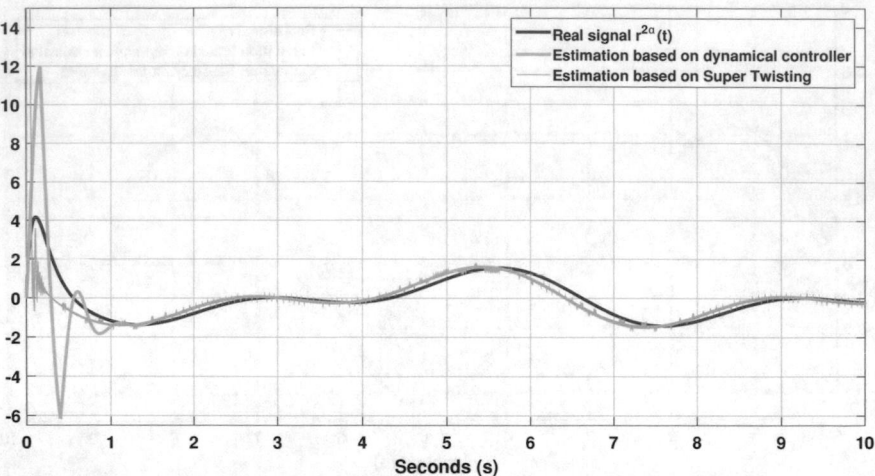

Fig. 9.5 Comparative analysis of the signal estimation $r^{(2\alpha)}(t)$ (noise free simulation)

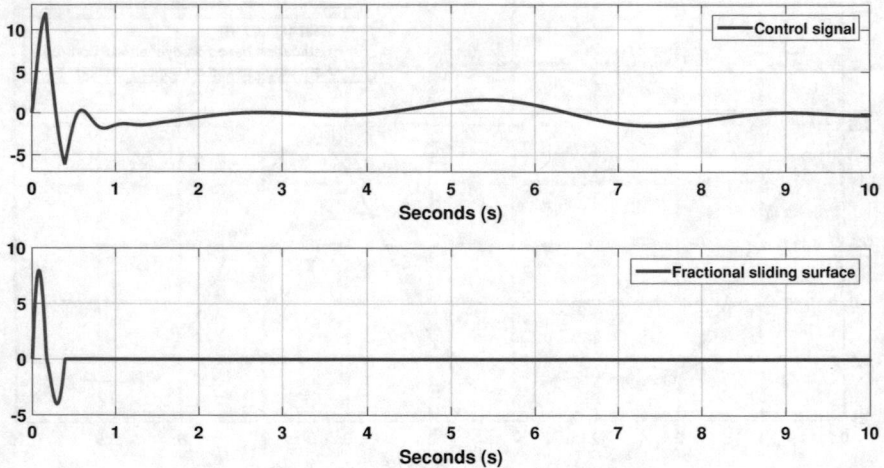

Fig. 9.6 Fractional control and fractional sliding surface (noise free simulation)

In Figs. 9.7, 9.8 and 9.9 the estimation of the signals are compared against the estimation using the FSTA. The proposed methodology shows an immunity behavior concerning the signal $r(t)$ when is contaminated with noise. The estimation using Ninteger Toolbox was performed without noise measurement with comparative purpose due to the lack of robustness when a noisy signal is considered.

Figure 9.10 displays the control signal $u(t)$ and the fractional order sliding surface when the signal $r(t)$ is contaminated with the Gaussian noise. The methodology illustrates how the fractional integration minimize the high-frequency noise and the chattering effect of the control $u(t)$.

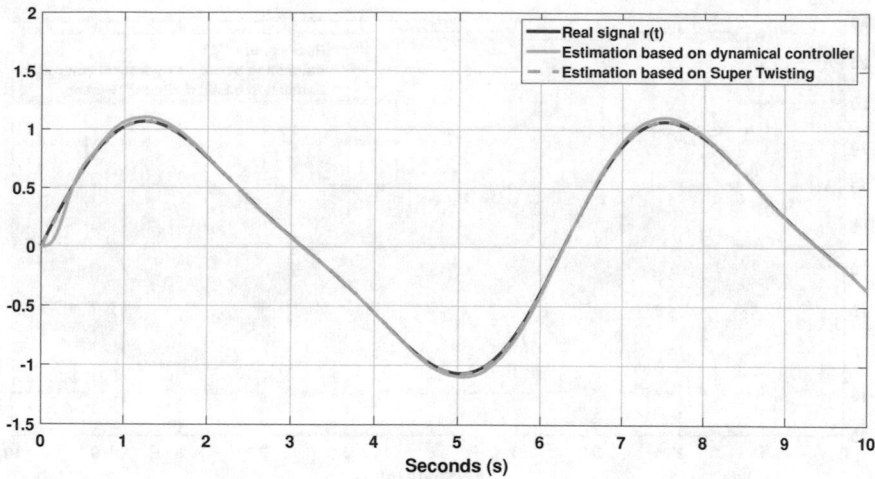

Fig. 9.7 Comparative analysis of the signal estimation $r(t)$ with noise measurement

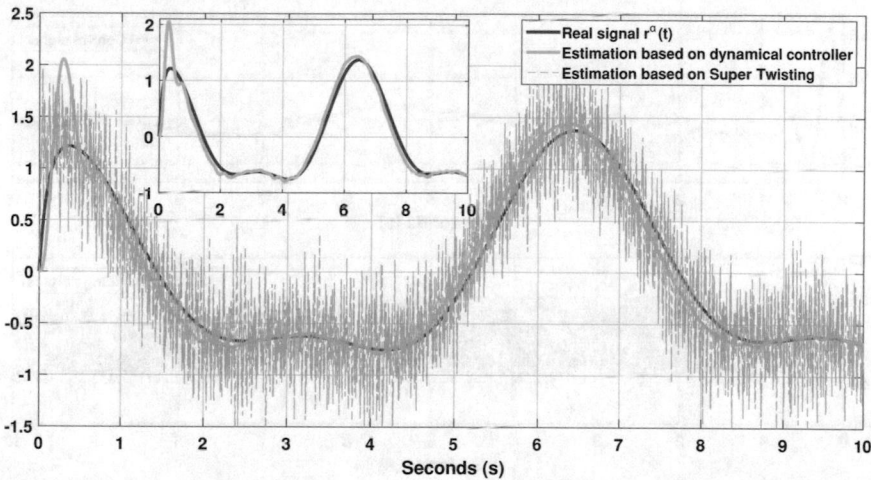

Fig. 9.8 Comparative analysis of the signal estimation $r^{(\alpha)}(t)$ with noise measurement

Finally, a performance index $J_i(t)$ is proposed with the aim of comparing in an objective manner the estimation of the fractional derivatives $^C\mathscr{D}^\alpha r(t)$ and $^C\mathscr{D}^{2\alpha} r(t)$. Both algorithms, the dynamic controller, and the fractional super-twisting are directly compared with the estimation based on the Ninteger toolbox.

$$J(t) = \frac{1}{t+\bar{\varepsilon}} \int_0^t |E_J(t)|^2 dt \qquad (9.39)$$

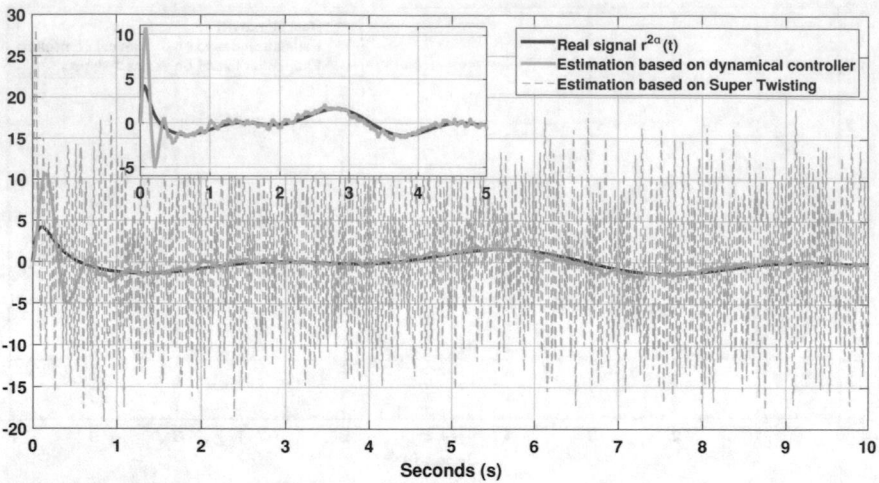

Fig. 9.9 Comparative analysis of the signal estimation $r^{(2\alpha)}(t)$ with noise measurement

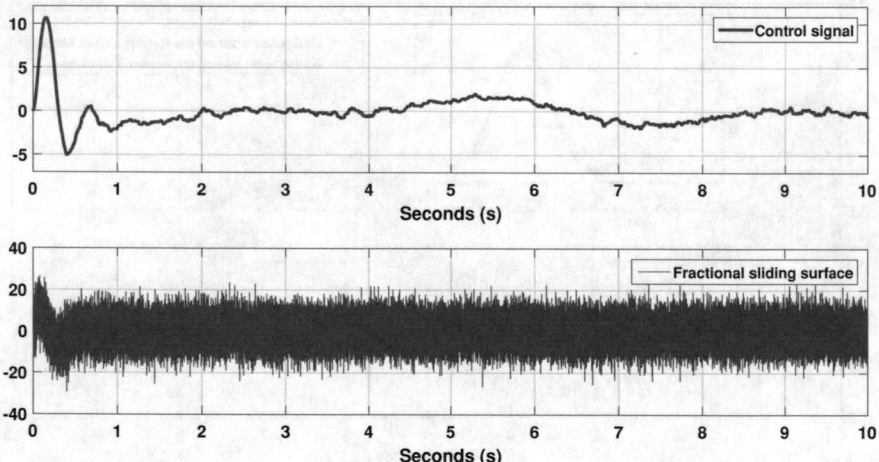

Fig. 9.10 Control signal $u(t)$ and fractional sliding surface considering a noisy signal $r(t) + \Delta_r$

where $\bar{\varepsilon} = 0.001$ and $E_J(t)$ is the performance error between the numerical estimation using Ninteger toolbox ($^C\mathscr{D}^{n\alpha}\hat{r}_N(t)$) and the dynamic controller ($^C\mathscr{D}^{n\alpha}\hat{r}_{DC}(t)$) or the FSTA ($^C\mathscr{D}^{n\alpha}\hat{r}_{ST}(t)$)).

$$E_J(t) = {}^C\mathscr{D}^{n\alpha}\hat{r}_N(t) - {}^C\mathscr{D}^{n\alpha}\hat{r}_{DC}(t)$$
$$or$$
$$E_J(t) = {}^C\mathscr{D}^{n\alpha}\hat{r}_N(t) - {}^C\mathscr{D}^{n\alpha}\hat{r}_{ST}(t)$$

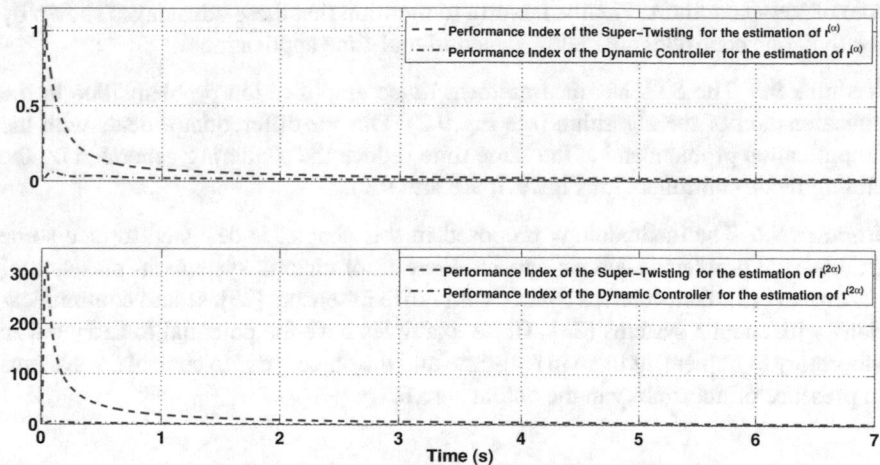

Fig. 9.11 Performance index for the estimation of $^{C}\mathscr{D}^{\alpha}\hat{r}(t)$ and $^{C}\mathscr{D}^{2\alpha}\hat{r}(t)$ considering a signal $r(t)$ without measurement noise

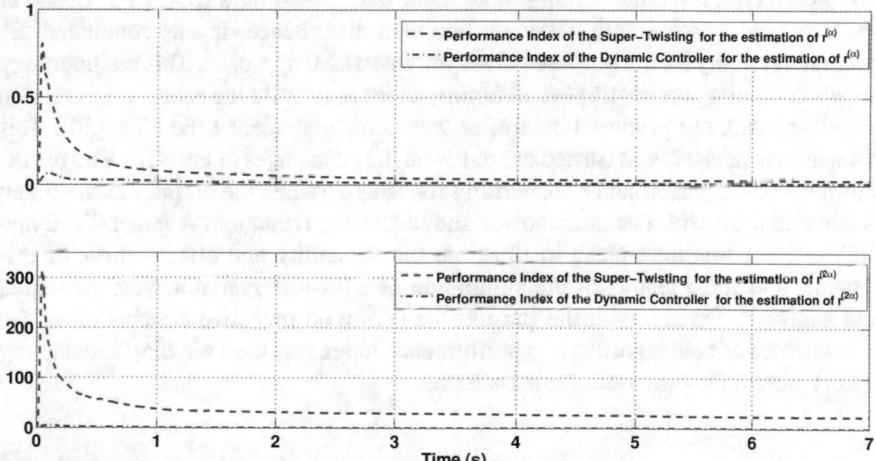

Fig. 9.12 Performance index for the estimation $^{C}\mathscr{D}^{\alpha}\hat{r}(t)$ and $^{C}\mathscr{D}^{2\alpha}\hat{r}(t)$ considering a signal $r(t)$ with measurement noise

with $n = 1, 2$.

Figures 9.11 and 9.12 indicate better performance in both cases (with and without noise measurement) when our proposed controller is applied for the estimation. Based on the first impression, it can be said that the dynamic controller offers in the transient state a smaller overshoot that the super-twisting. Moreover, Fig. 9.12 shows that the dynamic controller is robust against high-frequency noise since the structure of the proposed scheme is built by a chain of integrators which play the role of low-pass filters. Finally, it worth to mention that these advantages shown by the dynamic controller are highly valued in real-time applications.

Remark 9.5 The STA has measurement noise amplification problems due to the concatenation of the algorithm (see Fig. 9.2). Our $n\alpha$-differentiator deals with the amplification problem and at the same time reduce the chattering generated by the sliding mode controller (see Figs. 9.7, 9.8 and 9.9).

Remark 9.6 The methodology proposed in this chapter is designed to face some engineering problems such as: synchronization of chaotic systems in presence of uncertainty [5], fault-tolerant control with noise in sensors [25], secure communications with chaotic systems [23]. Those examples have the potential to carry out in laboratory to implement them in real-time and the objective is to control the systems in presence of uncertainty in the output signal.

9.5   Conclusions

We presented a fractional sliding mode controller based on a robust PI^α observer for fractional systems with unknown bounded disturbance. It was concluded $\mathscr{L}_2$ quadratic stability for our proposed observer via an LMI approach. This methodology allows computing optimally several tuning matrices to build the robust observer. On the other hand, our proposed controller was compared against the STA [30]. This comparative analysis was carried out to reveal the advantages of the STA. The results showed a better performance concerning robustness (when the output measurement is contaminated with Gaussian noise) and chattering reduction. A generalized $n\alpha$-differentiator was introduced to illustrate the versatility and effectiveness of this scheme, and some results of the simulation of a 2α-differentiator were presented and analyzed. Finally, with the purpose of giving an objective conclusion of the performance of both algorithms, a performance index was used yielding satisfactory results around the proposed control scheme.

References

1. Aghababa, M.P.: A novel terminal sliding mode controller for a class of non-autonomous fractional-order systems. Nonlinear Dyn. **73**(1–2), 679–688 (2013)
2. Aghababa, M.P.: Control of fractional-order systems using chatter-free sliding mode approach. J. Comput. Nonlinear Dyn. **9**(3), 031003 (2014)
3. Aguila-Camacho, N., Duarte-Mermoud, M.A., Gallegos, J.A.: Lyapunov func- tions for fractional order systems. Commun. Nonlinear Sci. Numer. Simul. **19**(9), 2951–2957 (2014)
4. Chen, D., Chen, Y., Xue, D.: Digital fractional order Savitzky-Golay differentiator. IEEE Trans. Circuits Syst. II: Express Briefs **58**(11), 758–762 (2011)
5. Cruz-Ancona, C.D., Martínez-Guerra, R.: Fractional dynamical controllers for generalized multi-synchronization of commensurate fractional order liouvillian chaotic systems. J. Frankl. Inst. **354**(7), 3054–3096 (2017)

6. Dadras, S., Dadras, S., Momeni, H.: Linear matrix inequality based fractional integral sliding-mode control of uncertain fractional-order nonlinear systems. J. Dyn. Syst. Meas. Control **139**(11), 111003 (2017)
7. Diethelm, K.: The mean value theorems and a nagumo-type uniqueness theorem for caputo's fractional calculus (corrected version) (2017), arXiv:1709.01113
8. Djeghali, N., Djennoune, S., Bettayeb, M., Ghanes, M., Barbot, J.-P.: Observation and sliding mode observer for nonlinear fractional-order system with unknown input. ISA Trans. **63**, 1–10 (2016)
9. Hamidian, H., Beheshti, M.T.: A robust fractional-order PID controller design based on active queue management for TCP network. Int. J. Syst. Sci. **49**(1), 211–216 (2018)
10. Herrmann, R.: Fractional calculus: an introduction for physicists. World Scientific, Singapore (2014)
11. Ibrir, S., Bettayeb, M.: New sufficient conditions for observer-based control of fractional-order uncertain systems. Automatica **59**, 216–223 (2015)
12. Jakovljević, B., Pisano, A., Rapaić, M., Usai, E.: On the sliding-mode control of fractional-order nonlinear uncertain dynamics. Int. J. Robust Nonlinear Control (2015)
13. Ji, Y., Qiu, J.: Stabilization of fractional-order singular uncertain systems. ISA Trans. **56**, 53–64 (2015)
14. Kamal, S., Raman, A., Bandyopadhyay, B.: Finite-time stabilization of fractional order uncertain chain of integrator: an integral sliding mode approach. IEEE Trans. Autom. Control **58**(6), 1597–1602 (2013)
15. Kilbas, A.A.A., Srivastava, H.M., Trujillo, J.J.: Theory and Applications of Fractional Differential Equations, vol. 204. Elsevier Science Limited, Amsterdam (2006)
16. Lan, Y.-H., Gu, H.-B., Chen, C.-X., Zhou, Y., Luo, Y.-P.: An indirect Lyapunov approach to the observer-based robust control for fractional-order complex dynamic networks. Neurocomputing **136**, 235–242 (2014)
17. Lan, Y.-H., Huang, H.-X., Zhou, Y.: Observer-based robust control of a $(1 \leq a < 2)$ fractional-order uncertain systems: a linear matrix inequality approach. IET Control Theory Appl. **6**(2), 229–234 (2012)
18. Li, C., Wang, J., Lu, J., Ge, Y.: Observer-based stabilisation of a class of fractional order non-linear systems for $0 < \alpha < 2$ case. IET Control Theory Appl. **8**(13), 1238–1246 (2014)
19. Li, Y., Chen, Y., Podlubny, I.: Stability of fractional-order nonlinear dynamic systems: Lyapunov direct method and generalized Mittag Leffler stability. Comput. Math. Appl. **59**(5), 1810–1821 (2010)
20. Liang, T., Chen, J., Zhao, H.: Robust stability region of fractional order PI^λ controller for fractional order interval plant. Int. J. Syst. Sci. **44**(9), 1762–1773 (2013)
21. Machado, J.T.: The effect of fractional order in variable structure control. Comput. Math. Appl. **64**(10), 3340–3350 (2012)
22. Mainardi, F.: Fractional Calculus and Waves in Linear Viscoelasticity: An Introduction to Mathematical Models. World Scientific, Singapore (2010)
23. Martínez-Guerra, R., García, J.J.M., Prieto, S.M.D.: Secure communications via synchronization of Liouvillian chaotic systems. J. Frankl. Inst. **353**(17), 4384–4399 (2016)
24. Martínez-Guerra, R., Gómez-Cortés, G.C., Pérez-Pinacho, C.A.: Synchronization of Integral and Fractional Order Chaotic Systems. A Differential Algebraic and Differential Geometric Approach. Springer, Berlin (2015)
25. Martínez-Guerra, R., Trejo-Zúñiga, I., Meléndez-Vázquez, F.: A dynamical controller with fault-tolerance: real-time experiments. J. Frankl. Inst. **354**(8), 3378–3404 (2017)
26. Meléndez-Vázquez, F., Martínez-Fuentes, O., Martínez-Guerra, R.: Fractional fault-tolerant dynamical controller for a class of commensurate-order fractional systems. Int. J. Syst. Sci. **49**(1), 196–210 (2018)
27. Oustaloup, A.: La commande CRONE: commande robuste d'ordre non entier. Hermes (1991)
28. Oustaloup, A., Moreau, X., Nouillant, M.: The CRONE suspension. Control Eng. Pract. **4**(8), 1101–1108 (1996)

29. Petras, I.: Fractional-Order Nonlinear Systems: Modeling, Analysis and Simulation. Springer Science and Business Media, Berlin (2011)
30. Pisano, A., Nessi, D., Usai, E., Rapaic, M.R.: Nonlinear discrete-time algorithm for fractional derivatives computation with application to $PI^\lambda D^\nu$ controller implementation. IFAC Proc. Vol. **46**(1), 887–892 (2013)
31. Podlubny, I.: Fractional Differential Equations: An Introduction to Fractional Derivatives, Fractional Differential Equations, to Methods of Their Solution and Some of Their Applications, vol. 198. Academic, New York (1998)
32. Sabatier, J., Agrawal, O.P., Machado, J.T.: Advances in Fractional Calculus, vol. 4, no. 9. Springer, Berlin (2007)
33. Samko, S.G., Kilbas, A.A., Marichev, O.I., et al.: Fractional Integrals and Derivatives. Theory and Applications. Gordon and Breach, Yverdon (1993)
34. Shah, P., Agashe, S.: Review of fractional PID controller. Mechatronics **38**, 29–41 (2016)
35. Shahri, E.S.A., Alfi, A., Machado, J.T.: Stabilization of fractional-order systems subject to saturation element using fractional dynamic output feedback sliding mode control. J. Comput. Nonlinear Dyn. **12**(3), 031014 (2017)
36. Shtessel, Y., Edwards, C., Fridman, L., Levant, A.: Sliding Mode Control and Observation. Springer, Berlin (2014)
37. Valerio, D.: Ninteger v. 2.3 fractional control toolbox for matlab. Lisbon Technical University (2005)
38. Victor, S., Melchior, P., Oustaloup, A.: Robust path tracking using flatness for fractional linear MIMO systems: a thermal application. Comput. Math. Appl. **59**(5), 1667–1678 (2010)
39. Vinagre, B.M., Calderón, A.J.: On fractional sliding mode control. In: Proceedings of the 7th Portuguese Conference on Automatic Control (controlo'06) (2006)
40. Wang, D., Zhang, R.: Design of distributed PID-type dynamic matrix controller for fractional-order systems. Int. J. Syst. Sci. 1–14 (2017)
41. Wei, X., Liu, D.-Y., Boutat, D.: A new model-based fractional order differentiator with application to fractional order PID controllers. In: 2015 IEEE 54th Annual Conference on Decision and Control (CDC), pp. 3718–3723 (2015)
42. Wei, Y., Karimi, H. R., Liang, S., Gao, Q., Wang, Y.: General output feedback stabilization for fractional order systems: an LMI approach. In: Abstract and Applied Analysis, vol. 2014 (2014)
43. Zheng, S., Li, W.: Stabilizing region of PD^ν controller for fractional order system with general interval uncertainties and an interval delay. J. Frankl. Inst. (2018)
44. Zhong, F., Li, H., Zhong, S.: State estimation based on fractional order sliding mode observer method for a class of uncertain fractional-order nonlinear systems. Signal Process. **127**, 168–184 (2016)
45. Zhong, Q., Zhong, F., Cheng, J., Li, H., Zhong, S.: State of charge estimation of lithium-ion batteries using fractional order sliding mode observer. ISA Trans. **66**, 448–459 (2017)
46. Zubair, M., Mughal, M.J., Naqvi, Q.A.: Electromagnetic Fields and Waves in Fractional Dimensional Space. Springer Science and Business Media, Berlin (2012)

Appendix

A.1   Appendix to Chap. 5

Laplace transforms of integer- and fractional-order functions and operators (Tables A.1, A.2, and A.3).

Table A.1 Laplace transforms of some basic functions

$f(t)$	$F(s)$	$f(t)$	$F(s)$
t^n	$\frac{n!}{s^{n+1}}$	$E_\alpha(-at^\alpha)$	$\frac{s^{\alpha-1}}{s^\alpha+a}$
t^α	$\frac{\Gamma(\alpha+1)}{s^{\alpha+1}}$	$t^\alpha E_{1,1+\alpha}(-at)$	$\frac{s^{-\alpha}}{s+a}$
e^{-at}	$\frac{1}{s+a}$	$t^{\alpha-1}F_{\alpha,\alpha}(-at^\alpha)$	$\frac{1}{s^\alpha+a}$
$t^n e^{-at}$	$\frac{n!}{(s+a)^{n+1}}$	$t^{\beta-1}E_{\alpha,\beta}(-at^\alpha)$	$\frac{s^{\alpha-\beta}}{s^\alpha+a}$

Table A.2 Laplace transforms of integral operators

$f(t)$	$F(s)$
$f(t) * g(t) = \int_0^t f(\tau)g(t-\tau)d\tau$ $= g(t) * f(t) = \int_0^t g(\tau)f(t-\tau)d\tau$	$F(s)G(s)$
$I^n f(t)$	$\frac{F(s)}{s^n} = s^{-n}F(s)$
$^{RL}I^\alpha f(t)$	$s^{-\alpha}F(s)$

R. Martínez-Guerra et al., *Fault-tolerant Control and Diagnosis for Integer
and Fractional-order Systems*, Studies in Systems, Decision and Control 328,
https://doi.org/10.1007/978-3-030-62094-3

Table A.3 Laplace transforms of differential operators

$f(t)$	$F(s)$
$D^n f(t)$	$s^n F(s) - s^{n-1} f(0) - s^{n-2} f'(0) \ldots$ $= s^n F(s) - \sum_{k=1}^{n} s^{n-k} f^{k-1}(0)$
$^{RL}D^\alpha f(t)$	$s^\alpha F(s) - \sum_{k=0}^{m-1} s^k \left[^{RL}D^{\alpha-k-1} f(t) \right]_{t=0}$
$^C D^\alpha f(t)$	$s^\alpha F(s) - \sum_{k=0}^{m-1} s^{\alpha-k-1} f^k(0)$

A.2   Appendix to Chap. 8

Simulink diagrams: See Figs. A.1 and A.2.

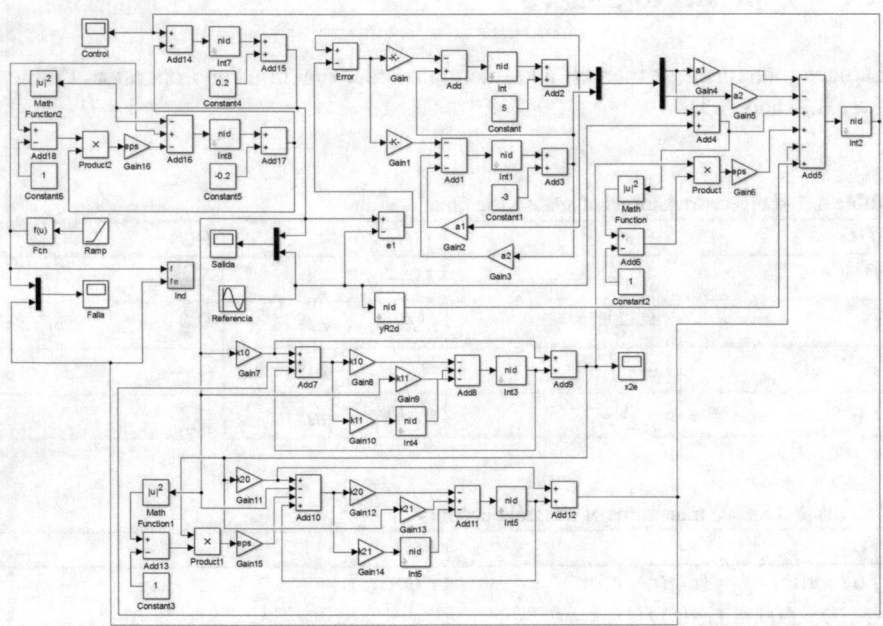

Fig. A.1 Closed-loop system of the Van der Pol oscillator (Matlab-Simulink®)

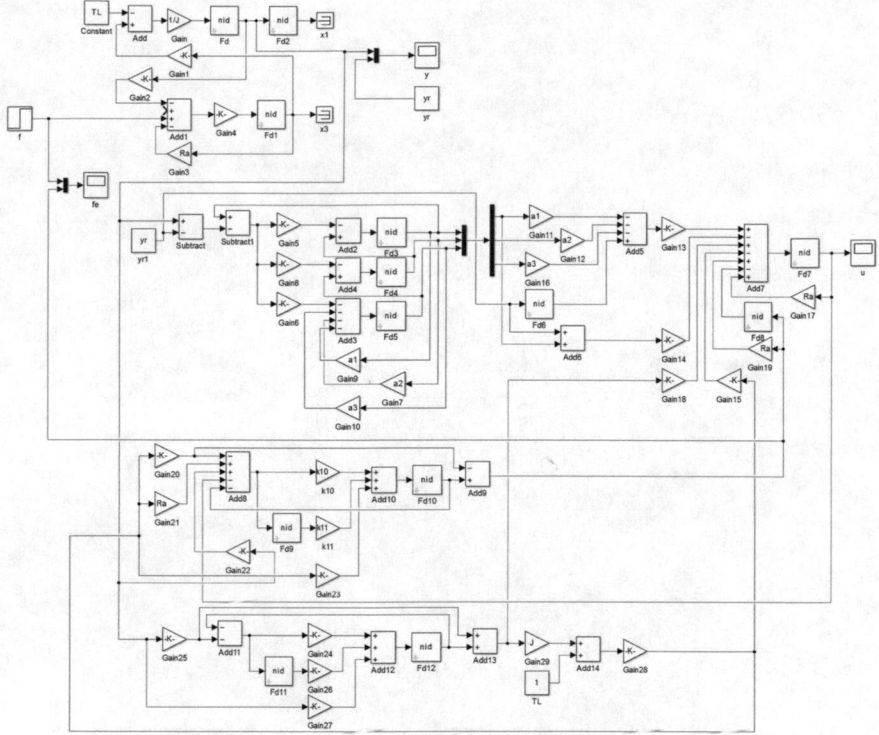

Fig. A.2 Closed-loop system of the DC motor (Matlab-Simulink®)

Index

© The Editor(s) (if applicable) and The Author(s), under exclusive license
to Springer Nature Switzerland AG 2021
R. Martínez-Guerra et al., *Fault-tolerant Control and Diagnosis for Integer
and Fractional-order Systems*, Studies in Systems, Decision and Control 328,
https://doi.org/10.1007/978-3-030-62094-3

Printed in the United States
by Baker & Taylor Publisher Services